SUBJECTIVISM IN ECONOMICS AND PHILOSOPHY

Re-orientating Economic Theory

Karl Mittermaier

Foreword by
Isabella Mittermaier

With a Preface by
Tony Lawson

and a Prologue by
Jochen Runde

BRISTOL
UNIVERSITY
PRESS

First published in Great Britain in 2025 by

Bristol University Press
University of Bristol
1–9 Old Park Hill
Bristol
BS2 8BB
UK
t: +44 (0)117 374 6645
e: bup-info@bristol.ac.uk

Details of international sales and distribution partners are available at bristoluniversitypress.co.uk

British Library Cataloguing in Publication Data
A catalogue record for this book is available from the British Library

ISBN 978-1-5292-5008-4 hardcover
ISBN 978-1-5292-4815-9 ePub
ISBN 978-1-5292-4816-6 ePdf

Cover design: Hayes Design and Advertising
Front cover image: Freepik
Bristol University Press uses environmentally responsible print partners.
Printed and bound in Great Britain by CPI Group (UK) Ltd, Croydon, CR0 4YY

FSC
www.fsc.org
MIX
Paper | Supporting responsible forestry
FSC® C013604

Bristol University Press' authorised representative in the European Union is:
Easy Access System Europe, Mustamäe tee 50, 10621 Tallinn, Estonia,
Email: gpsr.requests@easproject.com

For Isabella

Contents

Acknowledgements

Grateful thanks is extended to the two anonymous referees for their thoughtful discussions, recommendations and endorsements of this work for publication.

Permission to reproduce previously published material is acknowledged to the following publishers:

The *Cambridge Journal of Economics* is thanked for permission to include 'Menger's Aristotelianism' from its 2018 publication in Chapter 4, Section 5.

The *South African Journal of Economics* is thanked for its permission to reproduce '20th Century Strands of Austrian Economics' from Vol. 60(1), originally published as Section IV of 'Lachmann: A Biographical Sketch', now appearing as Appendix 1 to Chapter 4.

New York University Press is thanked for permission to include extracts from 'Mechanomorphism' in *Subjectivism, Intelligibility, and Economic Understanding* (1986) in Chapter 5, Section 8: 'Articulating subjectivist economics: from ontology to codification.'

Bristol University Press is acknowledged for permission to reproduce Section 10.3.1 from *The Hand Behind the Invisible Hand* (2020) in Chapter 5, Section 6: 'Intelligibility.'

The posthumous publication is greatly indebted to the referees and the publisher.

Foreword

Isabella Mittermaier

Prior to becoming an academic in May 1967, Karl worked for a financial services company preparing economic reports. He realized that apart from an understanding of certain institutional arrangements such as banking, the economics acquired in a formal education seemed to be of little relevance to the statistical investigations that are done. That was the starting point for this dissertation; a totally different approach to theoretical and empirical economic studies, and the fact that very little use seems to be made of economic theory in most empirical studies. There was a question to be investigated; what is the empirical content of economics.

This book was completed in its present form around 1981. Karl later drew on parts of it for separate publications and as reading material for his postgraduate courses. This text was conceived after Karl realized that his 1976 dissertation had been met with limited understanding. In response, the present text aims to make the earlier work more understandable, providing a philosophical context for the 1976 dissertation, now published by Bristol University Press as *A Realist Philosophy of Economics*.

When the 1976 dissertation was finished Karl asked me to accompany him when he went to hand it to a professor of economics at his home in Johannesburg. This puzzled me but I did go with him, and he did hand it to the professor. Periodically I would ask Karl what was happening about it and he always replied that one person did not understand it at all and the other person only understood 10 per cent of it. That is when he conceived of the text presented here as *Subjectivism in Economics and Philosophy*.

The book is the second section of Karl's 1976 dissertation. During the writing of which days and nights ran into each other. Karl grew a beard during that period and was never clean-shaven again. I typed the manuscript on a portable typewriter, transcribing Karl's pencil-written notes on foolscap pads.

Decades later, one morning roughly 12 years ago, I got up early and Karl, who had not gone to bed yet, was sitting on the staircase reading this dissertation and he said to me 'This is brilliant.' Karl did not blow his own trumpet.

I was never allowed into Karl's study. When I had to sort it out, I could not find the 1976 dissertation. I had sleepless nights thinking that I had failed Karl. Then I found the chapters, Michael Stettler, a former student and colleague of Karl's, and one of Karl's favourite young men, sorted out the chapters and collated some of them. Karl's other favourite young man, Eugene Le Roux, an accountant, spent his entire Christmas break putting it into digital format.

Karl's face would light up whenever Eugene or Michael came to visit him in his latter years.

Johannesburg,
January 2025

Preface

Tony Lawson
University of Cambridge

A book focusing on topics of lasting interest, written by one of the clearest and sharpest minds I have been fortunate enough to encounter.

Karl Mittermaier lectured in the history and the philosophy of economics at the Department of Economics at the University of the Witwatersrand, Johannesburg, South Africa. Mittermaier provided a critical and comprehensive account of economic thought and methodology using in particular the framework of the nominalism and realism divide, criticizing the former as it plays out in economic theory, inhibiting any clear understanding of economics and economic questions.

Mittermaier and I met somewhere in Cambridge about 30 years ago. My memory of details is hazy, but I do not forget the pleasure of the encounter. Mittermaier was clearly undogmatic, eclectic, innovative, challenging and interested in others. I well remember the unusual combination of really enjoying engaging with him, appreciating everything he said, and even agreeing on very much, whilst at the same time also disagreeing rather a lot on certain fundamentals. I think we were united in criticism, not least of mainstream economics, the absence of significant philosophical engagement in economics and the dominance of various positivist currents in approaches to methods that nevertheless prevailed. Our differences centred on our basic stances towards ontology and epistemology. In fact, Jochen Runde reminds me that when Mittermaier and I were introduced, Mittermaier informed me early on that his orientation was one of being ontologically cautious and epistemically bold, which was the mirror opposite of the emphasis of my own project on being ontologically bold and epistemically cautious. Prima facie such differences did seem rather fundamental. Yet I think that within the structure of these differences we found we shared an awful lot. Indeed, it felt strangely like we were members of the same research project.

Mittermaier was not a conventional academic. Although he wrote an enormous amount, he seems to have been reluctant to publish much of it.

Gradually, following his passing, some of his contributions are now seeing the light of day. And fortunately, the nature of its topics, much debated since the early Greeks, are such that Mittermaier contributions retain a lasting relevance. This current book exemplifies this feature. I read it with pleasure and profit despite its being written a while ago. In fact, for its insight and relevance, it could have been written yesterday.

Over the years Mittermaier was a deep influence on the local academic economics community in Johannesburg and many generations of students at Wits University. Now we can all benefit from his insights and wisdom.

14 November 2024

Prologue

Jochen Runde
University of Cambridge

Subjectivism in Economics and Philosophy is the final instalment of a remarkable posthumous trilogy by Karl Mittermaier. Written between his first and last books *A Realist Philosophy of Economics* and *The Hand Behind the Invisible Hand* (although published in reverse order) it fills an important gap in the record of his intellectual journey.

The topic Mittermaier addresses in this book is the relationship between theory and fact, or what he calls the problem of the general and the particular. The problem has occupied philosophers for centuries, and Mittermaier argues, the discussion often involved a shift from the 'natural' to the 'reflective' attitude, that is, from a consideration of events and things to a consideration of the terms in which events and things are understood, and the terms in which such understanding is itself understood. The aim of *Subjectivism in Economics and Philosophy* is to provide an account of the reflective attitude and, drawing on what were then more recent developments in philosophy and the philosophy of science, to flesh out a version for solving certain conceptual problems in economics. Mittermaier maintains that most of the criticisms levelled against mainstream economics by the Austrian school, for example, can be derived from the framework he proposes, and then in a way that makes it possible to go beyond criticism to constructive suggestions.

Like its predecessors in publication, *Subjectivism in Economics and Philosophy* stands testament to the rigour of Mittermaier's analysis, his erudition, and his almost unparalleled ability to mobilize the history of ideas to illuminate contemporary debates. But the three books are otherwise quite different. *Subjectivism in Economics and Philosophy* grew out of and is in some ways a meditation on his earlier *A Realist Philosophy of Economics*, a more narrowly focused critique of the foundations of mainstream price theory completed in the mid-1970s. The shift in thinking that the later book represents is due in large part to the influence of developments in postmodern philosophy associated with figures such as Husserl, Kuhn and Quine that hadn't entered the earlier work. Mittermaier uses the term 'theory-ladenness' as a catch-all term for the ideas involved, broadly that what people see is influenced by their presuppositions and especially beliefs to which they are firmly committed. This led him to a form of subjectivism focusing, not on the subjectivism of economic actors as exemplified in the work of G.L.S. Shackle and members of the Austrian school, but on the subjective perspective of the observing economist. The central idea is that what matters here is the coherence of what economists have to say about things. Mittermaier believed that such coherence is not to be found in the qualities or characteristics of the things themselves, but in how the things are perceived. Hence his interest in the presuppositions of the authors or schools he was engaging with, and his preoccupation with intelligibility rather than determinism as a suitable goal of economic theory.

While Mittermaier's first book can be seen as an instance of the more general approach subsequently developed in *Subjectivism in Economics and Philosophy*, the third, *The Hand Behind the Invisible Hand*, was written more or less in accordance with it. Unlike its predecessors, which were predominantly philosophical in orientation, this book is a contribution to political economy and political thought more broadly. Much of the argument takes the form of a critical survey of prominent champions of the market from Adam Smith onwards, looking at whether they viewed the rules and institutions required for the proper functioning of a market economy as something that themselves emerge spontaneously in the absence of government interference (the 'dogmatic' view) or as something that need to be deliberately created (the 'pragmatic' view that reserves an essential role for the eponymous, visible, hand behind the invisible hand). Following *Subjectivism in Economics and Philosophy*, the focus is on the presuppositions of his protagonists and, crucially, the practical consequences thereof: how dogmatic or pragmatic inclinations feed into views on the conduct of economic policy and the rationale of economic theory. Mittermaier here achieves the rare feat of someone working in the highly specialized area of the methodology of economics putting their own prescriptions to work in a follow up that will appeal to a much wider audience of economists and political scientists.

The publication of *Subjectivism in Economics and Philosophy* now makes it possible to appreciate the full sweep of Mittermaier's thought. Taken together, the three books represent an exceptional achievement, all the more so for someone who, apart from regular interactions with his mentor Ludwig Lachmann, was working pretty much in isolation. Mittermaier's diffidence and reluctance to publish were legendary amongst his colleagues in Johannesburg, and had he been more prepared to let things go he would surely have been one of the leading voices in the philosophy of economics of his generation. It is a tragedy that that this never happened and that he never received his due while alive. But at least we now have access to pretty much the full body of his writings. We all stand in debt to his widow Isabella and former student Michael Stettler for ensuring that it was published in its entirety and will survive to the benefit of future generations.

20 November 2024

Preamble: Re-orientating Economic Theory – Mittermaier's Challenge to Conventional Approaches

Michael Stettler
Department of Economics
University of Johannesburg

1. Introduction

Karl Mittermaier was sceptical about conventional economic theory. While working as an economist at a financial institution, where much of his work consisted of writing economic reports, he found that textbook economics had little relevance in compiling the reports. There was what he considered a disconnect between economic equilibrium theory and the answers he was expected to provide as an economist. It seemed to him 'that the mainstream of economic theory was formulated in such a way that it encouraged a neglect of the many points of orientation by which everyone reaches useful conclusions in his everyday economic life' (Author's Note, p xlii, this volume). This book, *Subjectivism in Economics and Philosophy: Re-orientating Economic Theory* and his earlier *A Realist Philosophy of Economics*, diagnoses the problem and suggests a solution. The type of empiricism he advocates as a solution is a re-orientation from the positivist type of empiricism to one more in accordance with Aristotle's empiricism.

Mittermaier argued – in *A Realist Philosophy of Economics* – that economic theory lacked empirical content. Although economics uses extensive statistical information, the data does not relate to theory in the manner thought to be the case. Others have also suggested that economic theory lacks empirical content (see, e.g., Hildenbrand 1999; Kirman 2021). Hildenbrand, for instance, argued that economics relies excessively on abstract models and artificial hypotheses, which are not borne out or validated by the data.

Mittermaier, however, shows that there is a fundamental misconception of theory and fact and how they relate. In the present text Mittermaier advances his argument, suggesting that the predicament stems from a misunderstanding of the relationship between the particular and the general (or universal) in theory.

Mittermaier's proposal for a re-orientation in economic theory is not merely a matter of philosophical preference or argumentation. His extensive discussions on how economic theory handles preferences and choice explain in detail how misguided the conventional approach has been. To put it briefly, in economic theory, the concept of choice has been redefined under the label of revealed preferences.[1] *Choice* has been given a new name; *preference*! This has gone largely unnoticed in the economic profession for 85 years, and its implications remain to be explored. One commentator says, 'A spectre is haunting economics – that spectre is revealed preference theory' (Beck 2024, p 288). Mittermaier has earned the right to propose a re-orientation in economic theory by uncovering the underlying issue regarding preference theory. This confusion comes at a cost, as is made evident by the recurring resurgence of debates surrounding this aspect of economic theory. He pinpoints the problem with revealed preference theory as arising from the priority given to the particular over the general, and it is here that the re-orientation must begin.

2. Argument summary

This chapter does not aim to summarize the various sections of Mittermaier's book; instead, it focuses on his central arguments in favour of re-orientating economic thought. Writing in the 1970s, Mittermaier demonstrated remarkable foresight in *A Realist Philosophy of Economics* and *Subjectivism in Economics and Philosophy* by identifying fundamental issues with preference theory in economics, highlighting the need for re-orientation. 'Revealed Preference theory has become a buzzword in philosophical debates about economics' (Beck 2024, p 289) and remains highly debated.[2] Some developments in economics, for instance behavioural economics and experimental economics, can be seen as reaction against revealed preference theory and assumptions about rationality in economics. Mittermaier's perspective, however, introduces new and unique views, as does his proposed philosophically motivated re-orientation or reversal.

Some have argued that in revealed preference theory *preferences* are given a technical meaning (see Thoma 2021). Mittermaier, however, identifies a different concern. The problem is not the technical definition of preferences but rather the rebranding of *choices* under the name *preferences*. This renaming disguises choices, which belong to the ex-post order of facts, as if they were ex-ante facts in the form of *preferences*. By doing so, the revealed preference theory creates a fundamental confusion between different orders of facts. This confusion stems from an inverted relationship between the particular (choices) and the general (preferences), misrepresenting the relationship between observed behaviour and underlying dispositions.

Section 4 highlights Mittermaier's bold methodological reversal in economic analysis. While conventional economics, rooted in an inductivist methodology, anchors general theories in particulars, Mittermaier, drawing on Aristotle and echoing Adam Smith, recommends everyday experience and its general meanings as a foundation and starting point for economic analysis, gradually refining these general meanings into distinct and precise frameworks, as he interprets Adam Smith to have done in *The Wealth of Nations*. Instead of validating the general through the particular, he advocates explaining the particular in terms of the general. To clarify this, Section 4 builds on the discussion in Section 3, which frames the debate by exploring Mittermaier's analysis of nominalism and realism – two contrasting approaches to understanding universal or general terms in theoretical analysis.

The relationship between the general and the particular is often discussed in terms of induction versus deduction or in the related terms of empiricism versus rationalism. This topic will be addressed in Section 5. However, Mittermaier considers the contrast between nominalism and realism more fundamental than that between empiricism and rationalism. Both rationalists and empiricists exemplify the nominalist position, treating universal theoretical terms not as abstractions anchored in reality but merely as expedients or labels intended to facilitate hypothesis testing. Both approaches presuppose a dualism in which words and facts are assumed to exist independently.

Some of the implications of a nominalist attitude towards words are explored in Sections 6 and 7. Once we hold that words and theoretical terms are not linked to the facts and particulars being observed but serve as mere labels and classificatory tools, selecting appropriate words takes priority over the meaning of words. More distinct words can serve more easily as labels. Distinct words are given priority status over other words. For instance, *norms* are somewhat indistinct, but the concept can be rendered more distinct when, for instance, '[s]ome have said that this is all there is to norms: they are tools of manipulation, used to dress up self-interest in a more presentable garb' Elster (1989, p 103). While this may render the idea of norms more distinct, it alters the word's meaning. Section 6 discusses Mittermaier's recommendation that the meaning of words be taken for granted instead of facts for granted, as the conventional approach does.

Thus, more fundamental than the particular–general distinction is the distinction, discussed in Section 7, between distinct words and those that are not. This distinction is the source of the differing methodological recommendations of Descartes and Aristotle. Descartes (2010 [1644], p 18) believes that 'for knowledge to be certain, the perception must be clear and distinct, and beyond all doubt'. Aristotle holds an opposing view, namely that we are more certain about that which is more 'confused' or more familiar.

The division between what is distinct and what is not, assumes an important role in Descartes favoured theoretical thinking and calls for an elaboration on the meaning of *distinct*. Paraphrasing Kant, one may say that '[a] concept is distinct if it is possible to determine its content with precision, i.e., if its boundaries are clearly defined' (Kant 1899 [1781], A241/B300), and that this is the notion that Descartes has of *distinct*. Aristotle holds, in contradistinction to Descartes, that we find more certainty in the more general and less distinct: I am more certain that a particular light in the sky is a celestial object than whether or not it is a planet, a star or a galaxy.

Given the two contrasting views of certainty and distinctness, there are two discernible approaches to attaining distinctness. These are discussed in Section 8, with the approach inherent in Aristotle and preferred by Mittermaier, referred to as the horizontal approach. This approach stays at the general or universal conceptual analysis level, progressing through this analysis to achieve a more distinct understanding of the concepts and their meanings. The vertical approach is associated with Descartes and the conventional approach in economics. It is considered vertical because it aims to descend from the theoretical term to the observable facts directly. The theoretical terms are selected for facilitating such a descent, which has become necessary because facts and terms are separated, namely the dualism Mittermaier identified as the commonality between the positivist approach and the Cartesian rationalism.

Section 9 addresses the question of which of the two approaches is the more superior. Mittermaier preferred the horizontal approach, but prominent scientists such as Heisenberg and de Broglie have also questioned the idea of clear and distinct ideas. More fundamentally, there are hidden assumptions in the argument that clear and distinct ideas coincide with certainty. A critical hidden assumption in Descartes' reasoning can be identified. Descartes assumes that because clarity, distinctness and certainty exist in mathematical knowledge, this relationship must automatically apply to all fields of knowledge, and therefore, in our economic thinking, we must select and begin with words that appear distinct. For Descartes, mathematics is a sign that what is clear and distinct must also be certain, and it is also the approach adopted in economics. But a sign is not an argument.

Section 10 discusses the conceptual framework that Mittermaier has developed, distinguishing between ex-post facts (like specific choices) and ex-ante facts (like enduring preferences), and reveals the hidden assumption that choice and preference belong to the same order of facts and are interchangeable concepts. This is what makes economic theory devoid of empirical content because its empirical content are records of past events (choices), but choices are not the empirical content of preferences, for preferences are in the ex-ante order of facts, whereas choices are in the ex-post.

Revealed preference theory attempts to derive general preferences from observed choices but conflates ex-post facts (past choices) with ex-ante concepts (underlying preferences). While Samuelson framed revealed preferences as a behaviourist alternative to utility, Mittermaier argues that the theory merely renames choices as preferences, creating an unjustified link between observable actions and unobservable dispositions. This highlights a nominalist approach and the persistent confusion in microeconomics between particular events and broader, general principles.

Section 11 concludes with a summary of the main points and a short introduction to what a system anthropinism, suggested by Mittermaier, might look like.

3. The universal and the particular in economic methodology: nominalist and realist perspectives

Economics, despite its abstract theorizing, claims to embrace an empirical methodology. The Greek origins of 'method' are 'meta' (over, etc.) and 'hodos' (way), giving us the meaning of methodology as the way towards knowledge. The empirical method progresses from particular empirical instances to general intellectual understanding and knowledge. Observational data is ostensibly given primary importance in economics, though this emphasis is often more rhetorical than real, as Deirdre McCloskey argued in *The Rhetoric of Economics*. Mittermaier's work is congruous with McCloskey's perspective – both were writing their respective texts around the same time – although he focuses on logical fallacies rather than rhetorical devices.

Mittermaier's work examines the philosophical tensions between theory and fact that underlie empirical methodology. He examines the 'universal–particular' or 'general–particular' distinction that connects abstract concepts to specific instances or theory to fact. We employ terms with varying degrees of universality – economy, law, utility, labour, preferences, capital, cost, competition, firm and individual – which often require clarification. Our standard model, rational choice theory, is typically framed in terms of cost-benefit analysis (actual dollars and cents) or supply and demand, into which these abstract terms are often incorporated.

What are these abstract terms, and what do they refer to? For example, Coase (1937, p 386) inquires into 'The Nature of the Firm', seeking 'a definition of a firm ... which is ... realistic in that it corresponds to what is meant by a firm in the real world'. Alchian and Demsetz (1972), however, argue that the firm is just a nexus of contracts, meaning that the firm is not a real thing; the real thing is the contract. *Firm* is a mere name for the epiphenomenon, so when we define 'firm', we do not define it as a real entity but as a name. In this view, contracts are assumed to be more real, and when we define 'contract', we can say that each contract defines itself by specifying what quantities must be supplied, at what price, and so on.

What we do when we define or use universal terms has implications for the research that gets done. For instance, Aristotle tries to define the word 'good'; depending on the definition he arrives at, it has normative implications for how one ought to behave to be virtuous. Others, like Hobbes, say that 'good' and 'evil' are labels we apply to things we like or dislike. This approach to 'good' is more compatible with positive economics because *good* can be identified in what individuals choose to buy or not buy, without normative implications.

It has implications for policy making. If we think a firm is just a nexus of contracts and contracts are just market bargaining outcomes (the market can be trusted to achieve the best outcome), we are less likely to consider regulating firms than if we have in mind some real definition to which firm behaviour has to correspond.

In Chapters 1 and 2, Mittermaier explores the two common responses to what has become known as the problem of universals: what do universal terms refer to? He explains: 'Realism and nominalism are the traditional competing theses on the relation between the general and the particular. … [R]ealism and nominalism differ on an ontological point, i.e. on the question of what is in the world. Realists hold that both particular and general terms refer to something real (hence realism)' (p 5, this volume).

Nominalism holds that only individual, concrete entities have existence. Consequently, universal terms, not being concrete entities themselves, function as conceptual labels, categorizing individual things into classes. Thus, according to Mittermaier (2020, p 79), '[n]ominalists … are committed to the view that the words which make up theories or hypotheses are mere names and are of no significance unless they are tied, rather like labels, to objects and events in the world or, as it became in positivism unless hypotheses are tested'. Chapter 2, 'The Problem of Similarity', can be illustrated with the example of the economic theory of addiction. Addiction may be seen as a real phenomenon or a descriptive label applied to similar behaviours. In common sense terms, addiction is often seen as a condition that causes people to act in specific ways, such as exhibiting price-inelastic demand for drugs. However, economic theory, particularly the perspective of Stigler and Becker (1977, p 81), reverses this causality and argues that 'addiction is the result of an inelastic demand for heroin, not, as commonly argued, the cause of an inelastic demand'. They contend that addiction is not the cause but the result of inelastic demand, meaning that the willingness to pay high prices defines addiction rather than addiction explaining the willingness to pay.

This theoretical inversion raises challenges for the nominalist perspective. Does the label 'addiction' capture an underlying essence or merely acknowledge superficial commonalities? The realist perspective holds that similar addictive behaviours share a common reality, whereas the nominalist view groups them based on similarity. The crucial question is: where does this similarity originate? Reducing addiction to a mere label strips the inquiry of its real-world significance. Are we arbitrarily defining the term 'addiction', or are we attempting to define the actual phenomenon constrained by the facts of the world?

Mittermaier introduces the distinction between nominalism and realism because it helps clarify central economic questions. For instance, he applied it to the prominent debate of the 1980s about the realism of assumptions. In particular, he analysed Friedman's 'positivist methodology in economics', concluding that 'positivism in economics is therefore a blend of nominalism and realism. Whether it is a blend which offers the best of both worlds is arguable. One reason for doubt is (to extend the metaphor) that it is a blend without fixed proportions, that everyone is his own blender' (see p 5, this volume).

The realist component of the blend is relegated to data selection. It is realistic because it presupposes that the words describing the data and the meaning of these words can be taken for granted. The nominalist element appears in Friedman's treatment of theory, specifically in how general concepts function as a filing system. As with any filing system that can be arranged at will, hypotheses can be developed without requiring realistic assumptions. What matters is whether the hypothesis predicts successfully. This nominalist inclination is compounded by a de facto positivism, where many economists see their task as devising hypotheses to fit facts, even if they don't explicitly defend positivist tenets. As a result, speculative models proliferate in economics, justified only by the hope that future testing will validate them (Chapter 2, Section 4, this volume).

> It is ironic that a philosophical movement, motivated from the start by an opposition to metaphysical speculation, should have had the effect in economics of encouraging a proliferation of speculative models justified only by a vague notion that obliging econometricians, presumably with direct access to theory-free data, will someday put them all to the test. (p 30, this volume)

4. Methodological reversal: from general to particular in economic theory

Mittermaier's work fundamentally challenges the conventional approach to economic theory, particularly its understanding of the relationship between general concepts (universal truths) and specific experiences (particular observations). He characterizes the conventional approach as taking the particular for granted in order to explain the general: 'We have seen that empiricists and positivists try to show how general propositions and their meaning are derived from or judged (verified, tested) by particular experiences. They want to explain the general and for this they take the particular for granted' (p 10, this volume). Instead, Mittermaier proposes a re-orientation that accounts for particular and observed economic phenomena through our general understanding. He calls it a 'reversal of what to many seems the normal procedure' (Chapter 5, Section 2, this volume).

He proposes a re-orientation in economic analysis, suggesting a shift from the standard approach of deriving general principles from particular events to one that moves from the general to the particular. This forms the core of his methodological argument. Mittermaier states:

> Here, however, the problem is being posed the other way around, in the way that it was posed by Aristotle and by Kant.
>
> It is in the particular event that is to be explained and for this one has to take meaning for granted. It is surely more sensible to show how the experience of particular events derives its meaning from prior conceptions. (p 134, this volume)

Mittermaier, to avoid confusion, wants us to distinguish between two meanings of particular: (1) the particular thing or event; and (2) the more particular as opposed to the more general concept. Referring to this second meaning of particular, Mittermaier (p 141, this volume) writes that 'in conventional discourse, there is no single criterion for distinguishing the particular from the general. Such discourse is partly based on the species/genus idea and the successively high degrees of abstraction of Aristotelian logic, and this makes the distinction largely a relative one'. For example, there are successively higher degrees of generality (or successively lower degrees of particularity) in the case of policies:

Interest Rate Policy → Monetary Policy → Macroeconomic Policy → Economic Policy

The other meaning of particular might be called the singular entity or event. Mittermaier (p 141, this volume; original emphasis) mentions the ostensive test 'that one can point to the particular', continuing that this '*particular is relative to the here and now of a perceiver*. On this criterion, a particular event or situation can be known to one individual' as opposed to a relatively more particular concept, which must still be known to more than one person for it to have meaning.[3] Throughout Mittermaier's text, the context will indicate whether the meaning employed is the meaning of particular as a particular thing or event, or the meaning of particular as a concept that is less general and abstract.

There are words where the different meanings of particular seem to converge. In economics that is the case, for instance, for the words *decisions* and *choice*. The word *choice* is particular in the sense that it refers to individual events that have occurred. However, *choice* is also particular in another sense: it is less general than terms such as *act* or *behaviour* or *preference*. Other words for which the two meanings of particular appear as if to converge (but of course such convergence is only apparent not real) include, for example, *cases*, *precedent*, *example*, *instance*.

So as not to confuse the two meanings of particular with regards to such words, one may notice that words have two aspects; the intension and the extension of a word. The extension of a word is all the things to which the word refers. The intension of a word refers to its meaning, or characteristics that define it. The extension of *choice* are all the acts of choice performed in the past or that will be performed in the future. Economic statistics are a compilation of choices (here the extension of the word) which have resulted in some exchange or transfer. Mittermaier refers to them as records of past events and as belonging to the ex-post order of facts. The intension of the word choice is its meaning as an act of selecting or deciding between two or more options. There is therefore always a gap between the singular particular and the concept particular.

Pareto (1935 [1916], p 57) grudgingly acknowledges these two aspects, when he writes that:

> A word designates a concept, and the concept may or may not correspond to a thing. But the correspondence, when it is there, cannot be perfect. Even if the word corresponds to a thing, it can never correspond to it exactly, in an absolute manner. It is always a question of a more or a less.

And as a nominalist, he will prefer the type of words where the correspondence between the word and the thing is close rather than far, namely, preferring words that are more distinct or determinate in the way defined above; a word is determinate[4] or distinct,[5] according to Kant's (1899 [1781], A241/B300) definition 'if it is possible to determine its content with precision, i.e., if its boundaries are clearly defined'. The word *choice* often appears to bridge the gap between the conceptual (an act of choosing) and the concrete (a record of a past event), such that thinking of general words as mere labels becomes plausible, where every individual choice could be given a unique number or date and time stamp to individualize it from all other acts of choice. Therefore words where the two meanings of 'particular' naturally converge, make them favourites for economists of the nominalist, positivist or behaviourist persuasion.

In economics, *choice* as a particular real event (ex-post fact) is a concrete observation, but naming it as *choice* or *preference* introduces a conceptual overlay that can shape interpretation. Economists must remain vigilant to avoid conflating the observable with the theoretical, ensuring that the act of categorization does not distort or overreach the evidence. Mittermaier's argument in *Subjectivism in Economics and Philosophy* and in *Realist Philosophy of Economics* is that they have not been vigilant in this regard.

As Mittermaier himself notes, his approach draws inspiration from the philosophical traditions of Aristotle and Kant. However, Aristotle's ideas, in particular, have faced intense criticism since the Enlightenment era, starting with thinkers like Descartes and Hobbes. The epistemological divide between Aristotle and Descartes lies at the heart of Mittermaier's proposal. What follows will elaborate on that divide while locating Mittermaier's approach within the divide, later justifying that approach.

5. Beyond the rationalism–empiricism divide: unveiling the nominalist connection

The Enlightenment challenge to Aristotle's pre-eminence has offered two alternative approaches, rationalism and empiricism, both in opposition to Aristotle's type of moderate realism. In *The Philosophy of Social Science*, Martin Hollis uses the analogy of spiders and bees, suggesting that the rationalist builds a network of concepts and theories that fit together logically like a spider spinning a web. In contrast, a more empirical approach, where knowledge is portrayed more like a beehive, is built from gathering external data.

It is commonplace to view the two approaches as being in conflict and competing, as exemplified by Hollis' very 'spider or bee' analogy. However, Mittermaier's explanation of the nominalist/realism divide shows that the rationalism versus empiricism debate is a red herring; they are two sides of the same coin, with both the rationalists and the empiricists subscribing to a nominalist approach to universal terms.

Rationalists, like Descartes and the British empiricists, particularly Hobbes and Locke, can be considered nominalists regarding their views on universals. Despite the differences, all three shared a nominalist inclination, rejecting the idea that universals are metaphysically real. Instead, each believed that our way of thinking – not something inherent in what they refer to – accounts for universals by which humans classify and understand particulars.[6]

The superficial dispute between rationalism and empiricism hides a much deeper congruence between the two approaches. This congruence is the shared view of universals or general terms as created rather than abstracted, and the commonality has facilitated the widespread adoption of the hypothetico-deductive method in economics, which integrates elements of rationalism and empiricism. The crucial distinction arises, however, in how they relate particulars to universals, with Mittermaier embracing Aristotle's approach over that of Descartes and empiricists.

Underlying both the positivist and Descartes' approaches is, according to Mittermaier (Chapter 3, section 5, this volume), a dualism they share. Positivism, particularly in its early forms, grounds knowledge in empirical facts by maintaining a Cartesian dualism between the objective, external world of observable facts and the subjective, internal realm of concepts shaped by human cognition.

> The elusiveness of the positivists' facts is inherent in the kind of dualism on which positivism is based. Though positivists treat the question of general and particular as a logical problem, there is in more popular versions of positivism (but not, for example, in the logical positivism of Carnap) a juxtaposing of an objective, external, real world and a subjective, internal conceptual realm unreal in the sense of being subject to human whim. (A development of Descartes's res extensa and res cogitans.) This dualism underlies the notion of facts used for testing hypotheses. As long as this conception is maintained, it will be necessary also to conceive the point of contact between the real world and the conceptual realm at which what is *given* by the real world is *taken* into the conceptual realm, at which facts are, so to say, handed over. (p 50, this volume; original emphasis)

Mittermiaer argues that the type of dualism is not feasible. The independence of facts (singular particular) from theory (general universal) cannot be maintained. The positivist attempt to separate 'given' facts from theoretical constructs is shown to be problematic. Likewise, constructing mechanistic/deterministic models (universal) for the human domain leaves the theory without genuine empirical content (particular), as it fails to capture the subjective nature of human action and understanding.

6. Taking meaning for granted

In Chapter 5 ('Anthropinism'), Mittermaier invites us to consider a starting point that is more fundamental than the general–particular distinction. He points to a more profound, often less obvious stratification influencing or shaping how the general–particular distinction is understood or applied. He suggests an approach that takes *meaning for granted* and focuses on how we articulate and understand economic experiences. 'What this amounts to is that we want to be able to conceive how particular events or changes are understood in terms of general and unchanging meaning' (p 133, this volume). In this text and *A Realist Philosophy of Economics*, Mittermaier frequently references and critiques revealed preference theory and Vilfredo Pareto's opinions, laying the groundwork for his proposed reversed approach. They serve as a counterpoint to his realist approach and his proposal to take meaning for granted and represent the prevailing perspectives he seeks to challenge. Throughout this discussion, they will provide a critical reference point.

For instance, Pareto condemned 'literary' economists as those who

> dilly-dally with speculations such as 'What is value?' 'What is capital'? They cannot get it into their heads that things are everything and words nothing, and that they may apply the terms 'value' and 'capital' to any blessed things they please, so only they be kind enough – they never are – to tell one precisely what those things are. (Pareto 1935, p 62)

Pareto's scepticism about indeterminate economic concepts is evident in his dismissal of value as a meaningful term. 'The term [value] has finished by designating some mystical, metaphysical entity or other that may mean anything since it has come to mean nothing at all' (Pareto 1935, p 30). Consistent in his critique, Pareto also rejected the concept of utility as similarly flawed, proposing instead the 'hypothesis of ophelimity' as an alternative (1935, p 29).

In contrast to Pareto's dismissive approach, Mittermaier emphasizes the importance of examining the meaning of economic terms. He explains that 'taking meaning for granted is equivalent to taking it for granted that people do articulate what is intelligible to them'. A sample of Pareto's own text illustrates Mittermaier's point, as Pareto assumes a shared understanding of various universals – such as truth, principles, science, experience, dogmas and inventions – without questioning their meaning or clarifying their significance, e.g. 'Experimental science has no dogmas, not even the dogma that experimental facts can only be explained by experience … and it in truth accepts the proposition that inventions may be promoted' (Pareto 1935 [1916], p 26). Irrespective of whether we are discussing general or particulars, and regardless of our preferred understanding of the general (nominalist or realist), meaning is taken for granted. Exceptionally, questioning the meaning of terms such as utility only proves the rule that we take meaning for granted. The questioning of terms, then, should serve as an occasion for clarification of the terms rather than their dismissal as metaphysical.

One may pointedly put it as follows: take facts for granted or meaning for granted. Mittermaier critiques the tendency to take facts for granted and proposes a reversal that prioritizes meaning over the particular or singular. Unlike Pareto and other empiricists or positivists, who purport to derive concepts like preferences and utility from observed behaviour, Mittermaier begins with meaning as the foundation.

Taking facts for granted, as is often done in economics, leads us to seek theories to impose meaning on the facts. However, this approach risks creating a disconnect between the theory and the facts. In revealed preference theory, for instance, 'choice' – the empirical data we observe – is presented as a manifestation of the broader concept of 'preference'. Yet, choice data merely record past decisions, serving as evidence of what was selected. While 'choice' and 'preference' share overlapping meanings, both grounded in the idea of selecting one option over another, it is not a species-genus relationship, where choice is an instance of preference[7] in the way that 'dog' is an instance of 'canine', nor therefore can choice be an instance of preference, in the way that Fido is an instance dog.

If every choice were a preference that is revealed, the meaning of preference would be contradictory to the very preference field idea that is at the heart of consumer theory. Consumer preference fields are assumed to be stable over a period of time and exhibiting consistency and intransitivity. But some choices are made 'on a whim', 'on the spur of the moment', 'on impulse', 'going with the flow', 'dictated by circumstances'; all notions that are incompatible with the meaning of preferences. For preference to have meaning it must be of a duration that extends beyond the moment of choice. A proper conceptual analysis of the terms is required, before the behavioural data as inductive evidence for preferences. It misses the fact that both choice and preferences are facts, but different kinds of facts.

Taking meaning for granted, as Mittermaier suggests, allows us to recognize a more fundamental distinction than that between general and particular: the division between (1) clear, distinct concepts amenable to quantification and (2) the more familiar, though less precise, concepts that form the basis of our tacit understanding. The one ends in quantification, and the other is in conceptual analysis.

This distinction between the familiar and the distinct underlies – and is more fundamental than – the distinction between the general and the particular. The familiar/distinct divide cuts across the general/particular dichotomy, such that both general and particular concepts can be either familiar or distinct. For example, the general idea of 'competition' is familiar, while the particular concept of 'perfect competition' is distinct but less familiar. The latter is an idealization of the former, where the idealization affords precision and distinctness, but the weight of meaning remains with the more familiar, more general concept. In other words, our understanding of perfect competition derives from, and is anchored in, our prior grasp of competition. The precision and clarity gained through idealization do not supplant the general concept, but it makes it capable of being mathematically modelled, though less realistic and relevant.

7. Distinct or familiar: two views of certainty

Mittermaier's (2023) extensive discussion of preference theory led to the conclusion that in economics, the term 'preference' – familiar but not very distinct – has essentially become synonymous with the more distinct 'choice'.[8] In common usage, choice and preference are obviously not the same. Economists justify this unusual usage, claiming it 'gives a technical meaning to preferences as they appear in standard economic theory, which consciously departs from the everyday understanding of preference as a conative mental attitude' (Thoma 2021, p 164).

In contradistinction to this permissive use of words, though supposedly technical and precise, Mittermaier demonstrates how this pursuit of technical precision in economics has paradoxically resulted in a more confused understanding. Distinctness of concepts and precision in language have been ideals in science in general and in economics in particular, going back at least as far as Descartes' expression of the ideal, but further still, for it is related to the nominalist perspective.

Methodologically, what is at stake here can be more fully brought out in terms of the opposing views held by Aristotle and Descartes. Aristotle is often contrasted with Descartes for being an empiricist. Aristotle's empiricism is evident in his assertion that the natural road of knowledge progresses from the senses to understanding. 'We do not understand a thing until we have perceived it.'[9] However, the contrast we are interested in here is more fundamental, relating to their respective views regarding familiar/distinct definitions of concepts and how certain we are in our knowledge of the empirical content of concepts, either more or less familiar or distinct. Descartes and Aristotle proposed contrasting methods for the pursuit of knowledge.

Descartes famously advocated for clarity and distinctness. In his *Discourse on Method* he had a list of four rules for scientific research: 'The first [rule] was never to accept anything as true if I did not know clearly that it is so … and to include nothing in my judgements apart from whatever appeared so clearly and distinctly to my mind that I had no opportunity to cast doubt on it' (Descartes 1637 [2003], p 50). Thus, not only did he advocate for clarity and distinctness in their own right, but he identified certainty and truth with that which is clear and distinct. It is probably not surprising that he would consider truth and certainty in this way, given his background in and contributions to mathematics and geometry, because in these subject areas, proofs establish truths that cannot be doubted. They are based on distinct concepts and logical steps. For instance, the statement that a triangle is a three-sided figure with interior angles totalling 180° relies on distinct numbers on distinct terms such as line, interior, angles, intersection and more. Geometry and mathematics, in other words, exemplify Descartes' idea that what is clear and distinct is also certain and that if we want to be certain, we must start with the clear and distinct.

Aristotle, conversely, argued that knowledge progresses from general, confused understanding to distinct comprehension.[10] For instance, we are more certain about identifying a plant as a tree than identifying it as a specific tree. We are more certain that banana trees are plants than trees. He saw an inverse relationship between immediate certainty and inherent knowability. In his *Physics*, Aristotle elaborated on this concept:

> Now what is to us plain and obvious at first is rather confused masses, the elements and principles of which become known to us later by analysis. Thus we must advance from generalities to particulars; for it is a whole that is best known to sense-perception, and a generality is a kind of whole, comprehending many things within it, like parts. Much the same thing happens in the relation of the name to the formula. A name, e.g. 'round', means vaguely a sort of whole: its definition analyses this into its particular senses. Similarly a child begins by calling all men 'father', and all women 'mother', but later on distinguishes each. (Aristotle 1990a, p 259)

This approach suggests that our initial understanding is often broad and imprecise, with particularity developing through further investigation. In the economic domain, for example, identifying competitive practices is generally easier than pinpointing specific monopolistic practices. Following two decades of relatively unchecked growth, the tech industry has become dominated by large corporations. In response, the European Union passed the Digital Markets Act (DMA) in September 2023, a landmark anti-trust legislation designed to counter the dominance of big tech firms and prevent monopolistic practices.[11]

In this context, Aristotle's use of 'confused' does not imply error but rather a lack of distinctness. He distinguishes between what is 'better known' (γνωριμώτερον) and what is 'clear' or 'distinct' (σαφής) in *Posterior Analytics* (Book I, Chapter 2, 71b33–72a5). This distinction highlights that we often have more certainty about general concepts than specific instances. For example, recognizing a tree is typically easier than identifying its exact species.

Mittermaier interprets Aristotle's perspective as emphasizing 'the fact that everything is known in terms of universals'. This interpretation contrasts with the Cartesian and positivist dualism of facts and theory. Instead, it aligns with 'the distinction drawn by Aristotle between primary and secondary substance or the unknowable "this" and the knowable "this such"' (p 50, this volume). This concept can be illustrated through the example of a bank loan.

Consider a specific loan. When asked, 'What is this?' we don't just identify it as 'this loan' but rather as 'a loan', a financial agreement involving borrowing and repayment with interest. This reflects Mittermaier's distinction between the unknowable 'this' (the individual instance) and the knowable 'this such' (the universal concept). Our understanding of economic concepts like loans is rooted in these universal categories, allowing us to comprehend various instances despite their differences.

In summary, we may name the two views of the relationship between certainty and distinctness: (1) the 'distinctness-certainty identity' and (2) the 'inverse certainty relation'. The first holds that what is most distinct is most certain and that certainty and distinctness go hand in hand. The second one holds that certainty is often inversely related to distinctness, and what is most familiar can easily be most certain for us, though not necessarily in a distinct manner; that certainty and distinctness may be inversely related. The word *good* is a good example. It is the most used adjective in the English language (Davies 2010) and while it is, therefore, the most familiar, it is nonetheless very indistinct, such that it is probably banished from all economics textbooks, while many more precise terms feature as proxies, such as maximization, optimization, efficient, productivity, etc. Mittermaier's reversal, therefore, is associated with a transition from the 'distinctness-certainty identity' to the 'inverse certainty relation' regards the expected relationship between distinctness and certainty, or as he puts it, reversing the neglect in economic theory of the familiar everyday economic life and experience:

> It seemed to me that the mainstream of economic theory was formulated in such a way that it encouraged a neglect of the many points of orientation by which everyone reaches practical conclusions in his everyday economic life, albeit unsystematic. An investigation of how economic theory had arisen out of everyday experience seemed to me of the greatest importance. (p xlii, this volume)

8. Cartesian versus Aristotelian approaches to economic knowledge: Mittermaier's perspective

Mittermaier advocates reversing economics' obsession with precise, distinct concepts (like 'revealed preferences') in favour of familiar but less distinct concepts that reflect everyday economic understanding, following Aristotle's view that certainty often comes from the familiar rather than Descartes' insistence that only clear and distinct ideas provide certainty.

The contrasting views of certainty have implications for research. For Descartes, clarity and distinctness are a starting point for further study. Rational choice theory in economics exemplifies this view, where rationality is clearly defined as making choices that maximize expected utility. Further research then concerns the particulars that validate the hypothesis. As this goes from theory to facts and facts to theory, this can be called vertical or top-down progression.

Mittermaier's reversal, instead, calls for a horizontal progression before a vertical movement.

The 'distinctness-certainty identity' suggests a vertical progression in knowledge, moving between the particular and the general. It is exemplified in the economic theory of revealed preferences, where the starting point is distinct choices – distinct in the sense of both distinctly observed choices and the clear and distinct concept of choice. From observed (or observable) choice behaviour, it arrives at more general inferred ordinal utility levels and preferences, namely the preference field.

In economics, vertical progression often proceeds in the opposite direction, such as when simplifying assumptions are made. For instance, assuming people maximize utility and observed behaviour validates this assumption. This is the case, for example, when it is argued in Becker and Stigler's *De gustibus* that even the drug has this utility when consuming increasing quantities of drugs since, given the depletion of his human capital, he now requires a greater amount of drugs. In other words, where common sense would say that a person's choices harm him, the Cartesian approach to the distinct-certain identity says that surprisingly/bizarrely, utility maximization is confirmed even in the most unlikely manner.

Or, when assumptions of perfect competition are made, and if the prediction of the model and actual observations bear out congruence, suggesting that, irrespective of the actual market structure, the results are 'as if' it were a perfectly competitive market. In all these instances, the point is that one moves to the particular without spending effort analysing concepts such as competition, preferences and choices.

The examples above are empiricist or behaviourist-positivist, but rationalist approaches also proceed in a similar vertical progression. Von Mises' a priori approach of praxeology moves from simple and clear definitions – that all behaviour is purposeful – to arriving at general conclusions valid at all times and everywhere, such as the law of demand holding, without much attention to explaining praxeology in detail, as Mittermaier notes:

> Perhaps because he conceived action as a Kantian category which all of us carry around with us all the time, he nowhere, as far as one can tell, tried to set out in a few propositions what praxeology entails apart from means and ends, i.e., he made no attempt to codify it. His discussion, in one of his books, on the logical character of praxeology was over and done within about 300 words. (p 149, this volume)

Descartes himself takes as a starting point his well-known 'Cogito, ergo sum' (I think, therefore I am), which he considers indubitable, implying that thinking and being are both clear and distinct concepts since what is certain must be clear and distinct, requiring no further clarification. Then, without further investigation of both the validity of his premise and further analysis of being, he develops distinct logical steps to arguments such that God exists.

On the other hand, Aristotle believes that clarity and distinctness are the endpoint of analysis, with the less distinct or confused and familiar being the beginning point. The other approach is exemplified by Aristotle's working out of the definition of motion, which seeks a clearer and distinct understanding of it ('Motion is the actuality of that which potentially is, insofar as it is potential'). Descartes rejected this 'metaphysical notion of motion, [as] is not intelligible' (Descartes 2010/1644), thinking instead that everyone knows what motion is.

Clearly, the textbook version of economics favours a Cartesian approach, readily moving from the concept to the quantitative. This may be referred to as the vertical movement, from the concept straight down to the statistics.

Mittermaier, on the other hand, values the Aristotelian type of horizontal movement at the same level of universality. This approach aims to clarify concepts without changing their level of abstraction. Mittermaier's *The Hand Behind the Invisible Hand*, exemplifies this approach. Unlike positivist economists, Mittermaier doesn't treat free market analysis as a direct description of economic reality reflected in statistics. Instead, he revisits Adam Smith's ideas, arguing that Smith's central question was how an economy should be structured to reward productive activities over predatory ones. Smith's *The Wealth of Nations* is best understood as an attempt to clarify the conditions necessary for achieving fair prices throughout the economy, *viz*. a 'system of natural liberty', characterized by free entry and exit in all markets.

Mittermaier emphasizes that Smith presented this as an intelligible ideal, not a description of existing conditions. The intelligible ideal – the system of natural liberty – articulates the meaning of 'fair prices', which serves as a heuristic device, where '[o]ne of its most important functions was to serve as a guiding conception for liberal economic policy' (Mittermaier 2020, p 73).

Mittermaier argues that the horizontal movement should precede the vertical movement in pursuing knowledge. He contends that one must first understand a concept distinctly at its general level before comprehending the basis for its division into particulars. This perspective contrasts with a modern tendency to move directly from the general to the particular without first achieving distinct knowledge at the general level. Such an approach, Mittermaier suggests, may lead to an inadequate understanding of the general concepts that underlie particular instances.

In summing the two approaches, Mittermaier emphasises that the one approach involves an analysis of a concept. The second approach implies the requirement for finding stable coefficients, that can be found (if at all) in the record of past events. The second approach of 'finding stable coefficients' is an instance of starting with a particular (or collection of particulars) and going from the particular to the general. One approach is the analysis of concepts, and the second is finding ways of handling and justifying them.

> When Smith recommended the dismantling of mercantilist policies and when Walras recommended what amounted to a programme

> of free-market pragmatism, their recommendations were based on analyses of the *concept* of free competition. The projected policy recommendations of those who pursue positive economics are based on the hope of finding stable coefficients by means of which economic affairs may be predicted and manipulated. (Mittermaier 2020, p 83; original emphasis)

9. Descartes' error: equating the distinct with certainty

The discussions surrounding Descartes' association of certainty with clarity and distinctness have persisted since its inception. British empiricists, beginning with the philosopher John Locke, have criticized Descartes' rationalism, which emphasizes reason, clear innate ideas and certainty over empirical evidence. However, there is agreement about their respective nominalism.

Critique from scientific perspective

Famous natural scientists of the 20th century have, however, found fault not so much with Descartes' rationalism as with his idea of distinct concepts, finding value instead in the more familiar though less precise and distinct conceptions. Louis de Broglie and Werner Heisenberg explain.

Louis de Broglie suggested that complete idealizations might be less applicable to reality. He proposed: 'We also could examine whether all idealisations are not that much less applicable to reality when they become more complete. … Though we have little inclination to be paradoxical, we could hold, contrary to Descartes, that nothing is more misleading than a clear and distinct idea' (De Broglie 1953, p 219).

Werner Heisenberg, pioneer of quantum mechanics and author of the uncertainty principle, literally revealed the limits of knowledge, proving that certainty is inversely proportional to precision. He argued against the overvaluation of precise rational terms, noting: 'The general trend of human thinking in the 19th century had been toward an increasing confidence in the scientific method and in precise rational terms. … Modern physics has in many ways increased scepticism, but it has at the same time turned it against the overestimation of precise scientific concepts' (Heisenberg 1958, p 201). Heisenberg further observed: 'one of the most important features of the development and analysis of modern physics is the experience that the concepts of the actual language, vaguely defined as they are, seemed to be more stable in the expansion of knowledge than the precise terms of scientific language' (Heisenberg 1958, p 200).

These scientific perspectives highlight the limitations of clear and distinct ideas in understanding reality, especially in the context of modern physics. They challenge the fundamental assumptions of Descartes' approach. For instance, his famous 'cogito ergo sum' (I think, therefore I am) may well be a statement that cannot be doubted, but that does not mean that we know clearly and distinctly what 'thinking' and 'existing' is.

Critique from logical analysis

More generally, Descartes' argument about certainty and clear and distinct ideas can be analysed as a third-figure enthymeme, a rhetorical device identified by Aristotle in *Rhetoric*. Descartes cites arithmetic and geometry as examples of clear and distinct knowledge. Mathematical concepts, such as 2 + 2 = 4 or a triangle's properties, are clear, distinct, and certain for us. Thus, he implies: (1) 'mathematics is certain' and (2) 'mathematics is clear and distinct', therefore 'what is certain is clear and distinct', but that is like arguing that: (1) Mittermaier is an economist; (2) he is German; (3) therefore, economists are German.

An enthymeme is a syllogism with an unstated premise. An enthymeme is arguing based on an assumption rather than something explicit. The unspoken premise is that what holds for mathematics and mathematical concepts holds for other subjects. This unstated premise allows Descartes to generalize from the specific case of mathematics to a universal principle about knowledge. However, this argument must be stronger, as it attempts to derive a universal principle from a particular case without adequate justification. This weakness in Descartes' approach leads to two problematic tendencies: assuming that certainty implies clear and distinct knowledge and assuming that clear and distinct ideas must be true.

If Descartes' position were correct and particular, distinct knowledge would be more precise and certain than general knowledge, leading to a radical shift in our approach to understanding the world. We could, in theory, abandon general knowledge entirely in favour of highly specific, particular knowledge. (Mittermaier will argue that this happened as an attempt in economic's revealed preference theory.) This would render broad concepts and generalizations obsolete, as they would be inferior in precision and certainty to more specific information. The fact that we cannot simply discard general knowledge in favour of particulars suggests that Aristotle's position more accurately reflects the nature of human understanding and the relationship between certainty and specificity in our knowledge.

10. The ex-ante/ex-post confusion in revealed preference theory

Revealed preference theory represents an attempt in economics to stay as true to the particulars as possible, to take the particular for granted and 'to show', as Mittermaier says, 'how general propositions and their meaning are derived from or judged (verified, tested) by particular experiences'. The theory, developed by Paul Samuelson in 1938 but building on Pareto's work, aims to derive individuals' preferences from their observed choices. Samuelson argued that consumer theory should not be based on axioms or general statements about preference and utility but on concrete cases – actual consumer behaviour. This procedure has been called behaviourist and is a species of positivism and empiricism, specifically the nominalist variant.

While revealed preference theory has established itself as amongst the most important theories in economics, for Mittermaier, it is a prime example of the confusion that arises from following what he called 'the normal procedure' of justifying the general through the particular. Contrary to its name, revealed preference theory is not about preferences but choices. Choice as a concept is determinate and distinct; the word's extension is choices performed, a subset of which have been compiled into economic statistics. The word's intension is its meaning as an act of agency. Choice, thus, is the ideal starting point for nominalist-inspired thinking, where each individual act can be envisaged as having been labelled.

Preference, on the other hand, is not a clear and distinct concept. Pareto would have called it metaphysical, just as he considered utility metaphysical. However, in the revealed preference theory, each choice reveals a preference. From a set of options, the consumer's indifference curve can be derived, which is a manifestation of the underlying preference field. In this manner, preference has been rendered precise and distinct; it can even be represented mathematically.

The apparent accomplishment of revealed preference theory is that choices are no longer just random records of past events but can be tied to or related to something more enduring, namely, a consumer's preference field. Individual choice is still particular but is now associated with something more general. Revealed preference theory might be considered the accomplishment of what Pareto (1935, p 16) thought to be the scientific method: 'While metaphysics proceeds from absolute principles to concrete cases, experimental science proceeds from concrete cases … to general principles.' Preference has become like the genus for choice, the species of revealed preferences. Unrevealed preferences are also a species of preferences that are unattainable, given the consumer's budget.

Mittermaier argues that revealed preference theory is based on a sleight of hand and that 'the normal procedure' aspiring to base or justify the general (preference fields) on prior observable particulars (choices) has not been achieved. In *A Realist Philosophy of Economics*, Mittermaier (2023) developed the conceptual framework of ex-ante and ex-post facts, which he then uses to show that in revealed preference theory, it is simply a matter of dressing up choices as if they were preferences.

Ex-post facts refer to past events, and ex-ante facts are facts or structures that are more enduring than the transient past events, in the way that the ability to feel hunger is more enduring than someone's last food choice. If an aircraft crashes, investigators will identify all manner of ex-post facts, trying to uncover the cause of the accident: pilot error, engine failure, weather conditions, explosion, etc. The ex-ante fact, on the other hand, is gravity. O'Donnell describes

> [e]x post facts [as] events that have occurred in the past prior to some moment in time. The subtlety is that this moment can either be now (in which case the facts have happened and are knowable) or in the future and that our direct experience, therefore, always relates to ex-post facts. (O'Donnell 2023, p xxvii)

> When he talks about the ex-post facts that one uses to construct explanations or stories to explain the past, he clearly denies the idea that we can extrapolate from that set of facts to predict the future. Why is that? It is because what happened is not only what results from the deterministic effects of certain causes but is also conditioned by chance and coincidence. (Kirman 2023, p xiv)

In the conceptual framework of ex-ante facts and ex-post facts, choices feature as ex-post facts. However, choices as ex-post facts are stochastic, representing a sequence of random events evolving over time. Pareto did not want to relate these concrete cases to utility and utility maximization, which he considered metaphysical, as that would have contradicted his notion of rational mechanics: 'The theory of economic science thus acquires the rigour of rational mechanics: it deduces its results from experience, without bringing in any metaphysical entity' (quoted in Mittermaier 2023, p 129). In place of utility, he proposed ophelimity, represented through the indifference curve. However, this concept carried too much unwanted meaning in the general, universal or metaphysical sense. His theory, he argued, should be based solely on facts, not universals that explain them. Ophelimity did not gain acceptance in the broader economics community.

Samuelson continued where Pareto had left off, hoping to free 'the theory of consumer behaviour … from any vestigial traces of the utility concept' (Samuelson 1938, p 61). One might be inclined to object that preference is a metaphysical concept as much as utility is, but that is to misunderstand what Samuelson had done. Revealed preferences are not put to work as a replacement for the concept utility. Revealed preferences perfectly track the actual consumer choices; they are the choices. In other words, Samuelson has given a different name for the concept choice, naming it 'revealed preference'. Rather than changing the name of the metaphysical entity (utility), Samuelson changed the name of the observable, creating the impression that the ex-post observable choices and the new name given (revealed preferences) were two different things. That revealed preference is just a new name given to choice is even admitted inadvertently by one of its proponents, when Binmore (2009, p 19) inverts our common-sense understanding of the matter: 'It isn't true that Pandora chooses b because she prefers b to a. On the contrary, it is because Pandora chooses b rather than a that we say Pandora prefers b to a.' In other words, preference does not refer to something real; it is just a name we give when we choose one option over another, thereby highlighting the nominalist attitude at work.

The fundamental flaw in revealed preference theory lies in its attempt to bridge an unbridgeable conceptual gap between ex-ante and ex-post facts. Choice can only be observed after it occurs – it is inherently ex-post. Samuelson's relabelling of ex-post choices as 'revealed preferences' implicitly suggests that these observations can inform us about ex-ante mental states or dispositions. However, this logical leap cannot be justified. Instead, as Mittermaier (2023, p 154) notes, all that has happened is that the ex-post choices 'appear … as ex-ante facts, in the guise of consistent preference fields'.

Preferences must exist prior to choice, if they are to have any meaningful predictive or explanatory power – they must be ex-ante. By simply relabelling 'observed choices' as 'revealed preferences', one cannot hope to establish a connection to genuine ex-ante preferences; it is a linguistic sleight of hand that obscures the distinction between what comes before choice (ex-ante) and what we can only observe after the fact (ex-post). Mittermaier states that the manner in which economics deals with preferences and choice is one that 'dresses up ex-post facts as ex-ante facts' (2023, p 18).

Revealed preference theory exemplifies 'how a confusion between ex-post and ex-ante facts has manifested itself in microeconomics' (Mittermaier 2023, p 18). This confusion arises from following Descartes' approach of starting with the clear and distinct – in this case, choice – then applying a label like preference, which carries connotations of greater generality and of ex-ante facts.

11. Conclusion

Mittermaier's critique of conventional economic theory, particularly revealed preference theory, exposes methodological problems in economics' treatment of the relationship between the particular or singular on the one hand and the general or universal on the other. His analysis shows how economics, pursuing scientific rigour through clear and distinct ideas, has confused different orders of facts and forced inappropriate relationships between concepts. The attempt to derive general principles from particular observations, exemplified in revealed preference theory's treatment of choices as preferences, demonstrates how economics has been led astray by its adherence to Cartesian ideals of knowledge and its nominalist treatment of universal terms.

Mittermaier's proposed solution, anthropinism, offers an alternative approach that begins with human meaning rather than theory-free particulars. By taking meaning for granted and recognizing the distinction between ex-ante and ex-post facts, this approach provides a more coherent framework for economic analysis. His re-orientation of economic methodology, drawing on Aristotelian insights about how knowledge progresses from familiar to distinct understanding, suggests a way forward that better captures the human-centred nature of economic phenomena.

Mittermaier's interpretation of Adam Smith's *Wealth of Nations* becomes particularly relevant in this context. In *The Hand Behind the Invisible Hand*, he portrays Smith's idea of a liberal economy as a working out of the implications of a free market and how this gives meaning to the concept of 'fair price'. Similarly, Smith's *Theory of Moral Sentiments* contains much that can be generalized into what Mittermaier called anthropinism.

The Theory of Moral Sentiments provides a historical precedent for an approach that takes human meaning as its starting point. In it, Smith develops what he calls the System of Sympathy, a portrayal of how individuals make judgements about the conduct of others. The System of Sympathy consists of (1) self-command, (2) sympathy and (3) the impersonal spectator. These three can be understood as instances of the more general human powers of (1) will, (2) intellect and (3) memory.

Economics has tried to reduce theory to taking account of choices and interesting larger economic phenomena in terms of individual choices, explaining the general using the particular. Mittermaier's reversal proposes to start with the general. Anthropinism would start with the most general of that which is unique to us humans, the three powers which constitute the system of anthropinism. Mittermaier uses anthropinism with overlapping connotations shared with subjectivism, particularly subjectivism in economics and philosophy.

Smith's system of sympathy has provided a careful distinction between virtue and self-interest. It is based on an understanding of human nature, where self-interest features as (1) self-interest as nature and (2) self-interest as will. By nature, we desire our happiness. But by human will, this desire can be directed to good or evil. Similarly, a system of anthropinism – economics that is not mechanistic but subjectivist – can accommodate economic conduct that seems incomprehensibly in equilibrium economics things, such as altruistic behaviour and other areas of paradigmatic puzzlement that have given occasion to research areas such as experimental and behavioural economics.

A system of anthropinism would free economics from confusing ex-post facts for ex-ante facts, starting instead with a general understanding but with analysis that will develop our knowledge to become distinct. With a distinct general understanding, such as also Smith's system of natural liberty, economics can then grasp and analyse facts, especially facts of the ex-ante order of facts, which will help us understand ex-post facts rather than the other way round, which is attempted by the conventional approach in economics and has led to confusion.

Notes

1 '[C]hoices … in the guise of consistent preference fields' (Mittermiaer 2023, p 154).

2 See, for instance, Beck (2024), Tipoe et al (2022), Thoma (2021), Guala (2019), Angner (2018), Clarke (2016).

3 In this context Mittermaier also distinguishes between a personal perspective and the impersonal perspective.

4 Determinate is used in regard to the intension of the word.

5 Distinct is used with regard to the extension of the word.

6 Mittermaier proposes metalanguage as an analytical tool to bridge nominalism's definitional precision with realism's conceptual richness. The framework distinguishes between observation language and theoretical language. The observation language adheres to nominalistic principles, making minimal ontological assumptions. The theoretical language, however, can incorporate Platonistic elements.

The idea of metalanguage has fizzled out because its rigid distinction between object language and metalanguage was challenged by post-structuralist and postmodernist thinkers, who argued that all language is self-referential and inherently meta, blurring the lines between levels of abstraction.

7 There would be at least two species of preferences; revealed preferences and unrevealed preferences, which are beyond the budgetary affordability.

8 More than that, to the more distinct concept, choice, is assigned another name normally referring to something other than choice (see Section 9, this chapter).

9 Aristotle (1990b: 97).

10 There is no contradiction between Aristotle's empirical assertion that knowledge progresses from senses to understanding ('We do not understand a thing until we have perceived it') and his theoretical methodology. Initially, understanding moves from particular to general, where 'particular' refers to perceivable singulars or ex-post facts (existing here and now, as described by Mittermaier). However, within theoretical contexts, 'particular' signifies not the singular but less general concepts, as seen in Aristotle's genus/species divisions (e.g. *firm* is more general than *oligopolist*). Notably, Mittermaier echoing Aristotle, recommends a reverse movement in theory, progressing from more general to more particular.

11 Specifically, the DMA targets six companies: Alphabet, Amazon, Apple, ByteDance, Meta and Microsoft.

References

Aristotle (1990a) 'Physics', *The Works of Aristotle: Vol. I.* In M.J. Adler (ed) *Great Books of the Western World, Vol.* 7, Chicago: Encyclopaedia Britannica, pp 259–358.

Aristotle (1990b) 'Posterior Analytics', *The Works of Aristotle: Vol. I.* In M.J. Adler (ed) *Great Books of the Western World, Vol.* 7, Chicago: Encyclopaedia Britannica, pp 97–142.

Beck, L. (2024) 'The Specter of Revealed Preference Theory', *Ergo* 11(10): 288–317.

Binmore, K. (2009) *Rational Decisions*, Princeton: Princeton University Press.

Clarke, C. (2016) 'Preferences and Positivist Methodology in Economics', *Philosophy of Science* 83: 192–212.

Coase, R.H. (1937) 'The Nature of the Firm', *Economica* 4(16): 386–405.

Davies, M. (2010) 'Exploring Linguistic Patterns in the Corpus of Contemporary American English (COCA)', *Journal of English Linguistics* 38(1): 3–19.

De Broglie, L. (1953) *The Revolution in Physics: A Non-Mathematical Survey of Quanta*, New York: Greenwood Press.

Descartes, R. (2003 [1637]) *Discourse on Method and Related Writings*, translated by D. Clarke. London: Penguin Books.

Descartes (2010 [1644]) *Principles of philosophy*, translated by John Veitch. Whitefish: Kessinger Publishing.

Elster, J. (1989) 'Social Norms and Economic Theory', *Journal of Economic Perspectives* 3(4): 99–117.

Guala, F. (2019) 'Preferences: Neither Behavioural nor Mental', *Economics and Philosophy* 35(3): 383–401.

Heisenberg, W. (1958) *Physics and Philosophy: The Revolution in Modern Science*, New York: Harper and Brothers.

Hildenbrand, W. (1999) 'On the Empirical Content of Economic Theories', in A.P. Kirman and L.-A. Gérard-Varet (eds), *Economics Beyond the Millennium*, Oxford: Oxford Academic.

Hollis, M. (1994) *The Philosophy of Social Science*, Cambridge: Cambridge University Press.

Kant, I. (1899 [1781]) *Critique of Pure Reason*, translated by J.M.D. Meiklejohn, New York: Dover Publications.

Kirman, A. (2021) 'Walras or Pareto: Who is to Blame for the State of Modern Economic Theory?', *Review of Political Economy* 33(2): 280–302.

Kirman, A. (2023) 'Extended Preface: A Realistic Attitude to the Economy', in K. Mittermaier, *A Realist Philosophy of Economics*, Bristol: Bristol University Press, pp xi–xxv.

Mittermaier, K. (2020) *The Hand Behind the Invisible Hand: Dogmatic and Pragmatic Attitudes on Free Markets*, Bristol: Bristol University Press.

Mittermaier, K. (2023) *A Realist Philosophy of Economics*, Bristol: Bristol University Press.

Mittermaier, K. (2025) *Subjectivism in Economics and Philosophy*, Bristol: Bristol University Press.

O'Donnell, R. (2023) 'Prologue: Mittermaier's Conceptual Framework', in K. Mittermaier (ed) *A Realist Philosophy of Economics*, Bristol: Bristol University Press, pp xxvi–xxxv.

Pareto, V. (1935 [1916]) *The Mind and Society. Vol. 1. Non-logical Conduct.* Edited by Arthur Livingston. Translated by Andrew Bongiorno and Arthur Livingston. New York: Harcourt, Brace and Company

Samuelson, P.A. (1938) 'A Note on the Pure Theory of Consumer's Behaviour', *Economica* 5(17): 61–71.
Stigler, G.J. and Becker, G.S. (1977) 'De gustibus non est disputandum', *American Economic Review* 67(2): 76–90.
Tipoe, E., Adams, A. and Crawford, I. (2022) 'Revealed Preference Analysis and Bounded Rationality', *Oxford Economic Papers* 74(2): 313–332.
Thoma, J. (2021) 'In Defence of Revealed Preference Theory', *Economics & Philosophy* 37: 163–187.

Author's Note

Before taking up a lecturing post at the university, I worked for five years as an economist for a large financial organization. During those years I got the impression that many members of the business community considered economists to be capable of doing things which in fact they could not do and that I was giving advice which in all honesty I was not equipped to give. It seemed to me that the mainstream of economic theory was formulated in such a way that it encouraged a neglect of the many points of orientation by which everyone reaches useful conclusions in his everyday economic life, albeit in an unsystematic way. An investigation of how economic theory had arisen out of everyday experience seemed to me of the greatest importance.

In my earlier book, *A Realist Philosophy of Economics*, I tried to make the point that the idea of *preference fields* in modern micro-economics, so to say, usurps the position of what might be called the empirical content of economics; more specifically that it makes it difficult to bring into consideration our everyday acquaintance with economic and social institutions. Further, I tried to show that the preference-field idea is a logical requirement once micro-economics was formulated as an axiom system interpreted as a deterministic system – a conception which at least in the early days of micro-economics was quite consciously based on the example of classical mechanics. The critique of this conception naturally involved some philosophical ideas.

After all this was completed, I wanted to find a firmer basis for the whole analysis. I embarked on an epistemological investigation, or rather a logical one, though not in the sense of formal logic.

The problem to which I now address myself, in the most general terms, is how the propositions of micro-economic theory relate to what is manifestly happening before one's eyes. Much can of course be said on this issue which, for short, I call the problem of the general and the particular. I consider some examples of how this issue has been treated throughout the centuries, drawing attention to a recurrent tendency, in Husserl-like phraseology, from the natural to the reflective attitude, i.e. one in which the terms in which events and things are understood, and the terms in which one speaks of such understanding, become the objects of attention without, however, any ontological claims being made for them. With this in mind, I try to develop a scheme whereby the problem of the general and the particular may be handled. Then, I try to show that most of the criticisms levelled against mainstream economics by the Austrian school may also be derived from this scheme which, however, also allows one to go beyond criticism to some constructive suggestions. My original point about *preference fields* then emerges as an example of this.

The scheme I have mentioned could probably be said to have an affinity with certain elements of Husserl's transcendental phenomenology and of Polanyi's philosophy and with certain remarks by the later Wittgenstein and the analytic philosophy he inspired. I do not claim to be *au fait* with these nor that the scheme falls squarely into any one of these genres. It is an ad hoc scheme for solving certain conceptual problems in economics. Here I present a general indication of what I mean by the reflective attitude. I do not want to express an opinion on whether it will contain anything of philosophical originality. I would claim some originality in applying a style of philosophical argument to economics. Logical positivism and Popper's ideas appear to have been the last philosophical movement to have had an impact on economics, though some economists nowadays occasionally speak of paradigms and research programmes in the fashion of Kuhn and Lakatos respectively.

Karl Mittermaier
Johannesburg
23 September 1980

1

Introduction

The theme throughout this investigation is the old question of fact and theory or of how the propositions of economic theory relate to what is manifestly happening before one's eyes. For the sake of brevity, this will be referred to as the relation between general and particular. However, all this should be seen in the context of discussions among economists. The mainstream of economic theory is dominated by equilibrium models. Sometimes these are regarded purely as axiomatic systems. Often, they are given a deterministic interpretation, i.e. they are taken to represent a process of determination in the economy – a conception which the founders of modern equilibrium theory consciously based on the example of mechanics. Often also the construction of equilibrium models is regarded as a matter of devising hypotheses to fit the facts, where it is not clear how the facts required for this purpose are to be conceived. Some schools of economic thought are extremely critical of this kind of economics. Members of one of these, known as the Austrian school, base their criticism on what they call subjectivism. They are critical of attempts by economists to copy concepts from the natural sciences, stress that economic action is planned in conditions of uncertainty (as it is put), insist on the unpredictability of human action, analyse economic phenomena in terms of the intentions of the people involved, and so on. Here it will be argued that the formal analysis of the relation between general and particular opens up a promising avenue of development for the economics of the Austrian school, but one that would put its contentions into a quite different perspective.

In order to facilitate a difficult exposition, the development of certain philosophical ideas will be traced from the preoccupation with permanence and change in antiquity to a style of philosophical reasoning which has established itself more recently. The analysis itself ultimately involves a careful articulation of certain ways in which we understand the term *fact*. When this analysis forms the basis of a critique of determinism and the preference-field idea in economics, the criticism differs from similar criticism by the Austrian school in so far as it may be said to be based on logical considerations rather than on an appeal to the apparent lack of realism of economic models. It is in this sense that the contentions of the Austrian school may be put into a new perspective. The analysis, however, need not be used only for, as it were, iconoclastic purposes but also for making constructive suggestions about the direction which the study of economics should take.

1. Philosophy of science applied to economics

Some preliminary remarks on methodology may be apposite. First of all, the word *methodology* poses problems. Some say that surely *method* is enough. Method sounds better, but one may object to the word *method* because it is not method that one ought to study. The word implies formulas, rules and routines, as though one could programme a PC to churn out journal articles. McCloskey (1983, 1985)[1] maintains that economists pay lip-service to certain rules of scientific method but that they don't really follow these rules and, moreover, that it is a very good thing that they don't. That far one may agree with McCloskey, but not when she seems to make out that the study of method or the philosophy of science is an attempt to create a kind of intellectual Gestapo that will suppress the free flowering of economic thought by insisting, for instance, on the testing of hypotheses. There is not much danger of that. If a person thinks he or she has a good idea, they will pursue it and try to make it known even if they do not follow the prescribed rules. Perhaps some ideas are suppressed; but that is not our real problem. The problem in economics is simply that there aren't many good new ideas about.

The approach here to the study of method, or the philosophy of science applied to economics (a term more preferable if it was not such a mouthful), is therefore not to report on the methods followed by great and wise men – such as Samuelson and Friedman. This may not be worth much. If it is to be worth doing at all, it must be done in an attempt to gain new insights that will allow one to do better economics. In a similar vein, it is a pity that economic methodology seems to be becoming a separate sub-discipline. Reflection on whether one's assertions make sense – and that is what is understood here by the philosophy of science – ought to be carried out by all economists. By making method a separate field of study, one is approaching the situation where one may ask an economist: 'Does your paper really make sense?' and receive the reply: 'Don't ask me. Methodology is not my field.' You may or may not agree that that would be ridiculous.

However, if, in reflecting on their work, economists are not constantly to reinvent the wheel, they ought to know just something, even if only in the broadest outline, about what has been said about some common problems over the centuries. The rest of this introductory chapter is therefore devoted to that end. First will be mentioned at least one reason why method is a difficult thing to study, taking the opportunity also of introducing a concept that will be useful later on. Then the present author shall be foolish enough to attempt to discuss a major, if not the major, theme of western philosophy. As it pertains to us, it is the question: What has economic theory to do with what is manifestly going on before our eyes, how is theory related to what we see when we walk down the road, go to a supermarket and so on. You may or may not agree that that question is sometimes a bother in economics.

2. Metalanguage: unit of significance from word to theory-laden facts

Apart from the awful jargon which infests this whole subject matter, perhaps the major difficulty of a study of method is that it is unavoidably metalinguistic. This requires one to explain briefly what a metalanguage is. Imagine that someone claims to know how the monetary system works and then spends 20 minutes talking about it. You now wish to comment on and criticize what the person has said. The subject of your remarks will not be the monetary system itself but someone's explanation of the monetary system. One says that the original explanation is in terms of the object language and the terms and propositions expressed in this language are then the objects discussed in the metalanguage. In other words, a metalanguage is a language used for talking about what is said in another language. Of course, the two languages are likely to consist of the same natural language, say English. Nevertheless, it is extremely important to distinguish between the object and the metalanguage. There is a world of difference between, for example, 'The cow jumped over the moon' and 'The little girl said "The cow jumped over the moon"'. You may be inclined to believe the one but not the other.

Logic necessarily and philosophy often is metalinguistic, and this makes the whole matter very tricky. It is very easy to get mixed up between the various levels of language and if you do that you will soon find yourself talking utter nonsense. There is another matter which we shall need in a moment. Since talk of method is metalinguistic, the objects talked about in discourses on method are terms, names, propositions and so on. A question then arises about which are the *units of significance* in such discourses. Traditionally, the significant units were terms and names (terms: e.g. capital, price, market; names: e.g. Mr Gorbachev, The NYSE). But then it was felt that these units were too small, and propositions became the significant units. (A proposition is what a statement or sentence is about.) This is still largely the case today, but in some interesting developments in the second half of the 20th century the significant unit became a complex or network of interrelated propositions, e.g. theories, paradigms, language games, and with it came the idea of theory-laden facts.

3. How theory relates to what is before our eyes

a. The one and the many or the general and the particular

We come now to the question of how theory relates to what is manifestly happening before our eyes. The ancient Greeks used to refer to this question rather cryptically as 'the one and the many', i.e. how one term can have many instances, how one theory can explain many actual cases. However, it will be better for us to consider the question under the heading: *The General and the Particular.* It is an enormous and difficult issue. But I have tried to find a shorthand way of giving a vague idea of what four well-known positions on the relation between the general and the particular, or between theories and facts, are about. Still, you might bear in mind the saying: Fools rush in where angels fear to tread. Except that this particular fool hardly ever rushes, the saying is apposite.

The particular at the level of the significant unit of a term is any proper name (such as the ones given above) or any reference to space and time co-ordinates, however vague (e.g. 28 September 1989 in the Barclays Building). A particular proposition is any proposition which contains at least one particular term. Theories, paradigms, etc., are never particular but they may have variables that take on particulars as values.

General terms are like the ones given above, capital, money, interest, human being, entrepreneur, and just about everything economic theorists (and theorists in other fields) talk and write about, and, for that matter, the majority of words we all use in the ordinary business of life. For general propositions we may borrow some notation from symbolic logic, namely: '(x): Ax implies Bx', where (x) is called the universal quantifier and A and B are general terms. The symbolism may be read as 'For every x, if x is A then x is B'; 'for every x, if x is human then x is mortal'; but also 'for every x, if x is a unicorn, then x looks like a horse and has a horn on its forehead'. For the last of these, one may add 'There are no unicorns' which a symbolic logician would express as 'For every x, x is not a unicorn'.

What is manifestly happening before our eyes, observations and reports of facts and events are all particular propositions. (The inverse, however, is not necessarily true. 'The rate of inflation in South Africa in August this year' is a particular but is not visible all at once to our eyes. But such particulars are derived ultimately from reports of observations.) If we now have another look at the example of a syllogism, we can see that the first line is a general proposition and the other two are particular propositions. The example represents the simplest case of what is meant by using a theory. A particular proposition leads via a general proposition to a conclusion which is also a particular proposition.

I hope I have said enough to give some idea of the distinction between the general and the particular, because all remaining remarks will deal with postulates about what the one has to do with the other.

b. The problem of induction

First to be dealt with is the *problem of induction*. It hails from the mind of David Hume, who was a friend of Adam Smith and something of an economist himself. The specie flow mechanism, for example, was first formulated by him. But Hume is most famous for having drawn attention to the problem of induction. Now, induction is the process of deriving a general proposition from a number of particular propositions, or the process of deriving a theory from the reports of a number of facts. The problem of induction is that there is no such process. That is to say, there is no rigorous, strictly logical way of making an induction. I may say: 'Name me anyone who lived 200 or more years ago and I say with confidence that that person is now dead.' Can I derive 'All men are mortal' from that? Strictly speaking, not. There may be someone alive today who will never die. Of course, we do in fact make inductions all the time but in the everyday context they are thought to depend on cultural and psychological factors. Some years ago, when I was writing something on this subject, I looked for an example of extreme inductive tardiness. The best I could find concerned a tribe in a remote part of New Guinea who apparently, at least when the anthropologists arrived, had never seen any connection between sexual intercourse and childbirth. In our own culture, we are much more apt to jump to conclusions. In scientific work there are, of course, conventions and approved methods for making inductions of which confidence intervals are an example.

From the point of view represented by the problem of induction, there is, therefore, strictly speaking, no connection between the general and the particular, or between theories and facts. Hume was what I am about to go on to explain, namely, a nominalist and a forerunner of positivism. He wanted to say, according to his precepts, that theories are merely mental expedients which have no foundation in logic but which we sometimes find useful in conducting our lives. Positivism and Popper's method of falsifying hypotheses in fact came up with what seemed a most ingenious way of side-stepping the problem of induction.

c. Realism and nominalism

Realism and *nominalism* are the traditional competing theses on the question of the relation between the general and the particular. Positivism is a form of nominalism and has for some time been the resident doctrine in economics to which, as McCloskey says, economists pay lip-service. Realism and nominalism differ on an ontological point, i.e. on a question of what there is in the world. Realists hold that both particular and general terms refer to something that is real (hence *realism*) whereas nominalists hold that only particular terms and the particular propositions in which they appear refer to something real that actually exists or happens. Particular terms are names which are tied to things and events rather like labels, but general terms and propositions are mere words, in the sense of hot air. This seemingly esoteric difference is nevertheless very important, especially in economics. It follows from the realist position that significant scientific conclusions may be drawn from the analysis of general terms (because they refer to something real) and you may or may not agree that economists quite commonly go in for such analysis. Nominalists, on the other hand, cannot maintain that anything worthwhile may be got from analysing hot air.

How then do realists and nominalists see the relation between the general and the particular? The realist position is usually illustrated with reference to geometrical figures. We may analyse the general term *circle* and come up with the theorem that the radii of a circle are of equal length. It may well be that the theorem does not hold of any particular circular thing like a wheel because it is not perfectly circular. But the theorem does tell us something about wheels and other circular things and, moreover, it gives us a clue about how to make more perfectly circular things. In other words, particular things approximate the general or the theoretical. (The problem that a miss may be as good as a mile has not to my knowledge been addressed by realists.)

The nominalist position as it has been worked out in positivism is known to economists mainly through Friedman's famous paper on 'The Methodology of Positive Economics', though Terence Hutchison's 1938 book called *The Significance and Basic Postulates of Economic Theory* is still the best exposition of positivism in economics. In short, the position is that hypotheses (general propositions) may be made up at will. But as hot air they are of no significance unless they are tested against facts (particular propositions). Such tests may be conducted if a hypothesis is so formulated that it logically precludes a particular. If then this particular is actually found, the hypothesis has been falsified otherwise it may be regarded as valid until such time that it may be falsified. This scheme cleverly side-stepped the problem of induction. While no number of particular propositions can prove a general proposition, only one particular proposition can disprove a general proposition.

Among well-known economists of the past, Menger was a realist, as was Walras[2] of sorts, while Pareto was proud to call himself a nominalist. Thereby hangs quite an interesting tale. In a few words: Walras' general equilibrium theory was an analysis of the idea of interrelated markets and of the general term *efficiency*, which, in analogy with what was said above about circles, was meant to tell us something of particular markets and market economies as well as to provide a clue about how to make the particulars approximate more closely to the general. Pareto, however, turned general equilibrium into a positivist hypothesis and thus into a kind of instrument for prediction. In time it was bound to occur to some economists that the theory would be a better instrument for prediction if it were reformulated to allow, for instance, for uncertainty and the costs of the dissemination of information. To Walras that would have seemed like saying that the theorem about the equal length of radii ought to be reformulated because there is not anyone in the world with a steady enough hand to draw a perfect circle.

d. Theory-laden facts

Finally, we turn to the notion of theory-laden facts. It has arisen over only the last 30 years or so and almost reverses the relation between general and particular that is presumed for the process of induction. Complexes of general propositions are held to affect particular propositions or, in other words, theories colour facts. Let me try to explain this rather difficult notion.

People educated in modern physical science have sometimes wondered how a belief in the efficacy of rain dances for producing rain can persist when rain dances obviously do not work. But to someone committed to a belief in rain dances that is not a problem at all. You perform a rain dance and then you see manifestly before your eyes that it is not raining and then you know that the gods are angry. One may ask: 'But what makes you think that the gods are angry?' 'Well obviously, it isn't raining, is it.' Quine, the American philosopher, has said that one can maintain a belief in just about anything if one is prepared to 'swell one's ontology' sufficiently, i.e. if one is prepared to impute the existence of sufficient entities to make the whole thing intelligible. Michael Polanyi, an academic chemist turned philosopher, has suggested that Freudians, Marxists and, he might have added, certain free market people preserve their commitment to certain beliefs in much the same way as the rain dancer. This is not meant to be deprecating. It now seems that the swelling of ontologies (i.e. the imputation of entities) is a fundamental aspect of human thought. Within one's complex of general propositions, i.e. within one's theories or paradigm, there are some to which one is so committed – Lakatos referred to a *core* – that one maintains them come what may. One may do that either by imputing particulars or by adding new general propositions to the complex or changing old ones to which one is less committed. Water at sea level is supposed to boil at 100 degrees centigrade. But if it does not, one is likely first to suppose that the thermometer is inaccurate, that the water contains impurities and so on. A rewriting of physics textbooks will be the last thing to come to mind.

But what has this to do with theory-laden facts? Well, let us consider whether the notion of a proper name – the mark of the particular – as a label tied to a thing is really satisfactory. Let me illustrate. A man walks into the room and my friend asks me: 'Who is that?' I reply: 'That's Zacharias Jones.' My friend says: 'Well alright, but who's Zacharias Jones?' It would not be very helpful if I pointed to the same man and said: 'That's Zacharias Jones.' My friend wants to know where the said Z.J. comes from, what he does for a living, why he is in the room and so on. A proper name is not so much a label on a thing as a label for a host of propositions containing the proper name. If that is so, reports of facts have deep roots in one's complexes of general propositions, one's theories or paradigm. The fact one notices will depend on the core propositions one is committed to. If different people have different commitments, they simply observe different facts. Facts are theory-laden.

This is of considerable importance in economics because there are so many schools of thought. One economist might say Zacharias Jones is an entrepreneur creating employment while another might say that Zacharias Jones is a capitalist exploiting workers. There may not be any way that the two economists could test their respective interpretations. They have different intellectual commitments and therefore may not have sufficient mutually agreed on facts to conduct the test. Theory choice, said Polanyi in his book *Personal Knowledge*, is ultimately an aesthetic matter. Western culture has had since the times of the ancient Greeks the intellectual value of simplicity. The simpler the explanation the more gratifying it is. Simplicity in this context is to be judged by how little imputation is required to achieve intelligibility and thus maintain one's intellectual commitments.

4. Plan of the following chapters

a. Chapter 2: The problem of similarity

Chapter 2 provides a brief introduction to, and historical survey of, certain issues in the theory of knowledge, reviewing some tendencies of thought on the nature of knowledge. An inventory of presuppositions and philosophical concepts is built up for later use. What can be inferred about the nature of things from, so to say, recurrent awareness is the (unsolved) problem of similarity.

With regards to positivism economics in particular, it will be suggested that economists who test hypotheses feel free to draw the line between the real and the contrived wherever they find it expedient to do so.

b. Chapter 3: General and particular judgements

This chapter provides a discussion of how the relation between general and particular has been handled in economics. The issue has been considered variously in terms of intelligible form and perceptible flux, quality and substance, the problem of induction, inductive logic of freely created definitions and hypotheses and the facts against which hypotheses are tested. In less philosophical terms the issue has been considered as the relation between intellectual disciplines such as economics and the objects and events of everyday life. The crude inductivism of earlier writers and the testing of hypotheses in more recent times are considered and reasons are given for not regarding them as adequate ways of treating the problem. An approach characteristic of some recent thought is explained in terms of what I call the subjective re-orientation of thought or a tendency to the reflective attitude.

We are of course interested in the matter only as a means to an end in economics, i.e. only in so far as it has a practical bearing on our understanding of the economy and economic events. (That it has such a bearing is also something we want to show.) For this purpose, the version which will be developed in Chapter 5 will suffice. We shall of course be able to draw on the ideas of various thinkers and especially on certain remarks made by Wittgenstein and Polanyi which seem to come to grips with the issues of transcendental subjectivism without a formal transcendental analysis. Nevertheless, the task of that chapter will not be altogether easy.

How inferences from such meditations can have a practical bearing on our understanding of the economy and economic events is a matter that will be argued in Chapter 5.

c. Chapter 4: Subjectivism in economics

The advent of subjective value theory a century ago created modern micro-economics. The notion of utility did not combine well with the mechanistic conception in economics because of its basis in the broad sense of pleasure and that this in principle proved to be the undoing of the ancient notion of utility in the mainstream of economic theory. Thus, the influence of the example of mechanics on the new theory in effect eliminated the subjective element.

The Austrian school did not develop in this way. This chapter will survey the methodological ideas of the leading figures of this school, demonstrating that in the Austrian tradition one finds subjectivism in two senses. One of these is related to the reflective attitude considered in Chapter 3. Menger's subjectivism of the subject has the clearest affinity with the re-orientation from the natural to the reflective attitude. On the argument of the previous chapter, this approach therefore should in principle bear the seed of a resolution of the problem of theory and fact in economics. It brought the distinction between general and particular into relief but the relation between them was not investigated. But it should also be clear from that chapter that Menger's treatment of the distinction between general and particular had a serious shortcoming.

d. Chapter 5: Anthropinism

Chapter 5 provides an outline of a system of thought, according to which we want to be able to conceive how particular events or changes are understood in terms of general and unchanging meaning. One should notice that the problem of knowledge is here not posed in what is nowadays the most usual way. We have seen that empiricists and positivists try to show how general propositions and their meaning are derived from or judged (verified, tested) by particular experiences. They want to explain the general and for this they take the particular for granted.

Notes

1 See Joseph Labia's (1988) review article on McCloskey's book *The Rhetoric of Economics*.

2 See also Clerq (1942) who considers Walras a realist.

2

The Problem of Similarity

1. The theory of knowledge

One cannot proceed much further with the inquiry begun in the previous chapter without formulating a more definitive conception of perceiving and knowing. In attempting such a formulation, however, one enters a well-trodden field. The issues with which the present-day theory of knowledge deals have been a major and growing preoccupation of western philosophic thought for some 2,500 years and it cannot be said that finality has been reached. Nevertheless, it is important that a stand be taken on them. Presuppositions about what we can know and how we can know it have a decided effect on the scientific work, especially when they remain largely unexamined. This chapter will give a brief review of some tendencies of thought on the nature of knowledge and build up, as it were, an inventory of presuppositions. The outline of a conception of perceiving and knowing will be attempted in Chapter 5.

It has often been pointed out that knowledge has a social aspect. One is apt to distinguish between fact and mere appearance or even illusion by whether others agree with one's perceptions. In academic and scientific circles, a new insight or a new finding has to pass through a whole social process of discussion and appraisal before it can become generally accepted as valid. The intersubjectivity of knowledge – the fact that it is shared by many – strengthens our belief that knowledge is of an objective reality. In one possible sense, therefore, the validity and objectivity of knowledge is a matter of its intersubjectivity, but the attempt to establish objectively in this sense whether there is such intersubjectivity would lead, in strict logic, to an infinite regress. Ultimately the individual has to ascertain for himself what there really is. Some knowledge, in the traditional view, he can confirm or establish simply by thinking. But for the other kind he needs, so to say, an empirical contact by which he ascertains that such and such is the case when it could well be otherwise.

How are we to understand such an empirical contact? In the ordinary business of life nothing seems more natural than simply to take for granted the objects and people, the events and situations of one's experience. On reflection one may perhaps suppose that representations of such things are somehow imprinted on the mind. But on further reflection one is likely to stir up a hornets' nest of conceptual problems. One may well imagine having a sense impression of a particular shade of red and one may suppose that it is a sensation stimulated directly by reflected light entering the eyes. But this does not take one far. Economics is hardly ever concerned with patches of colour, and it is not at all clear what could be meant by direct sense impressions of the things that do concern economists. Can one, for instance, have a direct sense impression of a price or of a market, or even of a capital good, or a borrower and lender concluding a financial transaction?

One way of treating these conceptual difficulties is to postulate that there are immediate sense data and that we interpret these, i.e. that we infer from these what people are doing, for what purposes things are used and so on. However, this leaves many questions unanswered. Where does the additional information come from, during the course of interpretation, if it is not in the sense data? Where do we draw the line between sense data and inferred entities? Do we actually see people and merely infer what they are doing, or do we see only colours and shapes which we first have to interpret as people before we can infer what they are doing?

In his well-known discussion on 'what is truth'[1] in economics, F.H. Knight took exception to positivistic views of interpretation. It may be instructive to quote from this discussion:

> Surely anyone who has made any progress at all in the study of philosophy, or even in private reflection about its problems, can be assumed to know that any simple antithesis between observation and inference is utterly untenable, if not downright foolish. The question as to the primary or immediate data of consciousness is perhaps the main perennial, unsolved and probably unsolvable problem of the theory of knowledge as a whole. (Knight, 1940, p 10)

He went on to say that it 'has been a commonplace, at least since the time of Kant, that ordinary *sense perception* is very largely an intellectual operation' and, as he had just told his economist readers, the 'heart of intellectual activity consists in the discernment of similarities and differences'.[2]

In the last of these extracts, Knight expresses what seems to be the most fundamental idea, in the theory of knowledge. The ability to discern similarities and the correlative differences seems to make knowing possible. When we are able to discern similarities, we are able to think in terms of types, of classes and of qualities or attributes. This in turn enables us to identify objects, relations, changes, etc., for to identify something, to know what it is, is to recognize it as of this or that type, to subsume it under a class or to ascribe certain qualities or attributes to it. When we establish empirically that this or that is happening or this or that situation prevails, we are identifying various elements of our experience and thereby both relating our experience to and setting it apart from others we have had.

Similarity is of course not totally unrelated to the notion of a sense impression, but it introduces considerations which remain unnoticed when one speaks of sense impressions. One should perhaps make a distinction between knowing and being aware – though no clear distinction is made in the everyday usage of these words – and associate the notion of a sense of impression with awareness. In the sense required one may well be aware of something that appears to be utterly unique, like nothing that has been experienced before. In particular, one's overall impression of each new situation one comes across may be one of uniqueness. But while one can be aware of uniqueness one cannot know it, one cannot articulate or describe it. For that one has to be able to identify things and one can do that only if one can discern similarities, i.e. if that of which one is aware is not unique or can be analysed into elements, some of which are not unique but recur frequently. What can be inferred about the nature of things from, so to say, recurrent awareness is the (unsolved) problem of similarity. But it is at least clear that it is not uniqueness as such that we can grasp intellectually. Uniqueness can be approximated in comprehension only as an ever-greater diversity of qualities or as an ever more distinct complex of attributes. Just as we are able to communicate thoughts because we can use a limited number of words to describe an infinite variety of circumstances, so we are able to know because we can discern similarities.

Before a distinction was made between science and philosophy, model builders were usually preoccupied with constructing cosmologies, speculations about the nature of the universe, and with ontologies of not necessarily visible and secular realms. The subject of knowledge was occasionally considered, even in antiquity – for example, in Plato's *Theaetetus* or in the teachings of Protagoras (man is the measure of all things, their existence or non-existence) which appear to have been not unlike much more recent doctrine – but it was not a major preoccupation. However, even in the construction of cosmologies the problem of similarity began to make itself felt, although often not explicitly. It led to speculation about universals or predicables (genera, species, qualities and relations that may be affirmed or denied of things and events), and this set the stage for a major re-orientation of philosophic thought, one which, after Descartes, was to confer ever more importance on epistemology or the theory of knowledge.

Knight's words may be made to serve here also to express the gist of this re-orientation. In the course of the discussion from which the previous quotations were taken, Knight said that 'we cannot separate the discussion of reality from the discussion of the knowledge of reality, the nature and structure of thinking and the conditions of its validity, or the workings of *mind* (meaning minds)'.[3] For cosmologists, knowledge (presumably including their own) was something that had to be fitted into the world, like everything else. Knowing was something that happened in the world. But when the new way of thought was developed, it began to appear much more as though the world is something that happens in knowing.[4] When we try to understand the world around us, we construct cosmologies or models of it, and we end up with exactly that – our *own models* of the world. It was a re-orientation of thought which Kant compared in importance with the Copernican Revolution. Henceforth there were also models of perceiving minds and of the acquisition of knowledge.

The new outlook also raised a new problem. It was now much more difficult to say what reality really was and what could be taken by the perceiving-mind as given by an outer reality. If the world one knows is merely one's model of it, then surely different minds, and especially different groups of minds, may know quite different worlds. Is it merely a metaphor to say that Marxists and neo-classical economists seem to talk of quite different worlds? Some, though only some, have been quite content to accept relativism, the doctrine that certain visions of the world are ultimate and that there are no criteria for determining their validity.[5] However, there is a further tendency for the new outlook to lead to solipsism, the idea that only the self is truly knowable and real, which has the implication that whatever else one thinks one knows is simply made up. This virtually no one has wanted to assert. Yet when the analysis is taken far enough it is surprisingly difficult to say just what it is that is *given*. On the other hand, our ordinary experience of life convinces us that there are *givens*, that, whether by the senses or by reason and reflection, we can have contact with the *external* or the *real* world (to use the expressions popular among philosophers and economists, respectively).

The following sections will give a brief outline of some of the more noteworthy contributions to the development of the ideas mentioned in this introduction.

2. Essence of change

The nowadays familiar features of the problem of knowledge began to emerge in antiquity when some ancient Greek thinkers turned away from the mythological explanations of the origin of the universe which were then current. The earliest speculations were attempts to find the basic substance or the substratum of all things. It was a feature of these cosmologies, as of all thought at the time, that qualities were never conceived as entities distinct from things. The hot, the cold, the dry and the wet were what we would now call physical things. Everything that existed and could be talked about was corporeal in the sense that it was tangible, had size, had a definite location and so on. This was not explicitly asserted, of course – it could not have been, since the distinctions necessary for this had not yet been made. In other words, it was not materialism as it is understood nowadays, namely, that only matter is real and all else mere appearance, because the separation of matter from quality, mind, spirit and so on had not yet been carried out.[6] The limitations of this way of thought were soon brought out by the questioning attitude which was then becoming characteristic of the Greeks.

They were apparent in the aphoristic statements of Heraclitus of Ephesus, who quite possibly found the language of his time inadequate for expressing his thoughts. It was the genius of Heraclitus, as Sir Karl Popper has put it, 'who discovered the idea of change'. 'The view he introduced was ... that the world was not a more or less stable structure, but rather one colossal process; that it was not the sum-total of all *things*, but rather the totality of all events, or changes, or facts' (italics in the original). Popper continued: 'The greatness of this discovery can hardly be overrated. It has been described as a terrifying one.'[7] The question which imposed itself on Heraclitus was: What does exist if, on the evidence of the senses, everything is forever becoming something else? His most famous saying was that you cannot step twice into the same river (though there is a scholarly dispute about the actual wording).[8] In a consistent corporealist conception, a river must be equated with flowing water and then of course it never remains the same. Yet, though no specific substance corresponds to it, one does speak, and presumably did then, of the same river. This kind of reasoning led to the classical problem of being and becoming or of essence and change. It has significance even for present-day economics, because economics, more obviously than the physical sciences, is concerned with events rather than with things.

It would be wrong to say that Heraclitus found no place for being and structure in his view of the world. He did see a substratum. It is usually translated as fire, but it seemed to signify something for which there is no modern counterpart. There was an interesting side to this substratum, namely the *Logos*, a kind of underlying formula of change or law of becoming, which saw to it that everything had its measure. Heraclitus was preoccupied with strife between opposites (thought of as constituents of things, rather as we may say that hydrogen and oxygen are the constituents of water); the hot becomes colder, the cold warmer, and so on. Everything was seen in terms of such strife. In the case of relatively stable objects, the strife is a tension between opposites and in the case of the visible flux, it is a continuous cycle of transformations. But no opposite can ever win in this strife, because an opposite cannot exist by itself (i.e. in modern terms, they are correlatives). To see this was for Heraclitus to begin to understand how the *Logos* manifests itself in a kind of dynamic equilibrium of opposites and how this constitutes the whole *Kosmos*. Furthermore, to understand the *Logos* was also to see the essential unity of opposites (in modern terms, perhaps temperature instead of hot and cold, but expressed picturesquely as, e.g., that the tension in a bow string is both a pulling apart and a pulling together). For this he even alluded to perspectives. The same sea water is wholesome to fish but detrimental to men. The way up and the way down are the same way (i.e. they differ only with respect to viewpoint). From their particular perspectives men see only the strife between opposites and the continual flux. But if they understood the *Logos*, they would see not only the equilibrium among all things but also their essential unity. Although it was conceived as a constituent of corporeal things rather than as something brought out by analysis, the Logos-Fire of Heraclitus certainly came close to the conception of incorporeal being.[9]

Some years later, being and becoming were seen in quite a different light by Parmenides of Elea. In a manner somewhat akin to the modern axiomatic method, Parmenides, starting with the premise *esti* – it is or exists – logically derived the nature of being, of an intelligible rather than sensed reality, and posited it as the only truly knowable. It was, however, the corollary of this that made a great impact. Parmenides had discovered the deductive method and he deduced from the idea of corporeal being that becoming, passing away, change and motion were impossible. If all that which exists once came into being and will one day perish, from what did it arise and where will it go to? He thought that these notions involved the belief that what *is* nevertheless *is not* and vice versa. If a thing is coming to be or ceasing to be, it both is and is not yet or no longer. If a thing is changing into another, it is and yet is neither the one nor the other. In fact, there could not be two things for then each would be and also would be not the other.[10] What it perhaps amounted to was that Parmenides had realized that logical truth is timeless. It does not come to be, nor will it cease to be true. It does not change or move about but is the same everywhere.

For anyone accustomed, as we are nowadays, to a distinction between assertions (which may be true or false) and an independently existing reality, his arguments are unlikely to be convincing. But to Parmenides and his disciples of the Eleatic School the assertable or the intelligible *was* reality. They actually believed that their senses deluded them, that the perceptible world was an illusion and that knowledge of it (the cosmologies of their predecessors) consisted merely of deceptively ordered words attached to appearances. This remarkable faith in discursive reason, even when the conclusions contradicted common-sense experience, fascinated the ancient world and affected virtually all subsequent thought. When, some time later, Leucippus and Democritus put forward their atomic theory, it was in deference to Parmenides that atoms were conceived as ungenerated and indestructible; and it was in order to restore the possibility of change and motion that the atomic theory (atoms moving in space) was based on the explicit paradox that the non-existent does exist, because empty space to the Eleatic way of thinking was the non-existent.[11]

The school of thought and religious brotherhood founded by Pythagoras had by this time developed the idea that all existence was to be accounted for by numbers and geometrical structure. The initial inspiration had been the discovery (probably by Pythagoras himself) that the concordant musical intervals corresponded to simple numerical ratios. Taking harmony as their paradigm, the Pythagoreans extended the numerical and geometrical analysis to all else. The parallel with the conceptions of modern physics has often been remarked upon, but it is of course not at all a close parallel. To the Pythagoreans mathematics had a mystic-religious significance – numbers ruled the cosmos. Furthermore, they were not conscious of a distinction between abstract and concrete. According to later commentators, they believed that (what we would now call) physical objects as well as moral qualities and ideals actually were, or were made up of, numbers, though sometimes their mode of expression apparently suggested that things imitated numbers or that the spirit of number entered into things.[12] In the terminology which was soon to be evolved, it could be said that they stressed form rather than matter, though they were not actually conscious of the distinction. This was the most important aspect of the influence they exerted.

The moral philosophy of Socrates introduced the moral form. Since Socrates wanted to counter the relativism on questions of morality which was coming into vogue during his time, he stressed the independence and immutability of the moral form. Courage, justice, etc. were constant, eternal ideals to which men could aspire in their actions. It was in this intellectual setting that Plato formulated his doctrine of Ideas from Eleatic and Pythagorean sources and that a belief in incorporeal being was explicitly stated. The Platonic Ideas or forms were not perceptible but intelligible. They constituted an intelligible world which was certain and immutable and outside of space and time. It could be reached by thought and reason alone and was in sharp contrast with the restless welter of the perceptible world in which nothing remained the same for long and therefore nothing could be known for sure. The connection between the two worlds found expression in the famous allegory of the cave.[13] Those who put their trust in the senses are like prisoners in a cave who have been shackled since childhood in such a way that they face the far wall and cannot turn their heads. A light casts shadows of the goings on behind them onto the wall in front of them. Never having seen the light or the objects behind them, they can do no better than to assume that the shadows are the only reality there is.

Plato had reached a position which in a sense is the reverse of that in which economists have found themselves in recent years. To us the ever-changing world of sense perception is the real world and the models which are meant to make it intelligible to us are thought merely to approximate this world. To Plato the intelligible forms constituted the real world, and the world of sense perception merely approximated it. A ring and a wheel would have been similar to Plato because both were imperfect manifestations of the circle, an eternal form which was not a construct of the mind but had a real, non-material existence. Change in this scheme was a coming to or ceasing to simulate or participate in one or more of the eternal forms. Plato was preoccupied with mathematical and moral concepts (not unrelated in his view) and in these fields his arguments were, and to many still are, quite plausible. It is far less convincing to say that individual men are imperfect manifestations of an actually existing Idea of man, or that a carpenter in making a table aspires to the Idea of tableness.

Aristotle was critical of Plato's conception of Ideas and brought his more empirical inclination to bear on the subject. He did share some of Plato's and Parmenides' misgivings about the perceptible – for instance, he ascribed the scepticism of some of his predecessors to the fact that they identified knowledge with sensation and existence with the perceptible flux – but he reconciled these views with an empirical method. He undertook to investigate substance, by which he meant individual, particular being or anything which is a *this*, as he put it – the individual man, the individual horse were his examples. He accepted that substance is unique and cannot be predicated (affirmed or denied) of anything else. In fact, substance is that of which universals are predicated. Knowledge, he pointed out, arises out of predication (a premise also reflected in his syllogistic logic). As he put it, we can never know a *this* but only a *this such*. In other words, when we ask: 'What is this?' the answer can only be: 'It is a such.' Again, when we ask: 'What is the just or the beautiful?' the answer can only be: 'It is a such and such.' To Aristotle such answers would indicate that something is a member of a particular species, which in turn is a member of a particular genus, and so forth. While what he called primary substance is therefore unknowable, substance as an object of thought consists of species and their successive genera.

The nature of a species – that which it really is or is said to be – Aristotle called its essence and individual members of a species have an essence in common but matter of their own. Essences are articulated as definitions which must ultimately be based on observation and reflection. To know the essence of something is to know to what species it belongs. It may be differentiated from other members of the species by predicates which attach to it not essentially but accidentally. (E.g., The tree on the hill and the tree in the valley are trees in essence but differ in accidentals.) In what may nowadays be described as the traditional view, it is accidental features that are left out when essential features are abstracted.

Having made up his mind about being, Aristotle still had to deal with becoming and change, for post-Parmedian thought was such that this question could not be ignored. His analysis is complex and turns on his distinction between being actually and being potentially, i.e. being capable of effecting, undergoing or resisting change. Nevertheless, one aspect of his analysis should be mentioned. In a spirit of inquiry one may ask of something either what it is or how it came to be. The first question asks for the essence of a thing, the second for its causes. Aristotle listed four types of causes, only one of which corresponds at all closely to present-day concepts of causation. In the case of, say, an earthenware vase, the potter and his instruments are the efficient cause of the vase; the raw clay out of which it is made, its material cause; its shape or form, its formal cause; and the purpose for which it is made, or the end sought (e.g. to display flowers, to earn the potter a living), its final cause. A thing can therefore be understood in terms of the interplay of various causes, and this brought Aristotle back to essences. These can in many cases be defined in terms of causes. The essences of artefacts, in particular, can usually be stated in terms of final causes. For example, a house is something that is meant to provide shelter, etc. The essence of a heap of rubble left by a builder, on the other hand, could perhaps be stated in terms of efficient causes.[14]

Aristotle did not pursue this line of thought far. He merely wanted to show that essences could not exist apart from other aspects of substance. They were *in* phenomena, as it was put later. This was to him an important point because he constantly criticized Plato for conceiving universals as independently existing entities, a conception which he thought led to all kinds of logical difficulties. There is, however, some doubt whether Plato really held the Ideas doctrine as dogmatically as Aristotle made out.[15] Moreover, essences still presuppose the meanings of cause and of the various universals that enter definitions. These were the *basic premises* of knowledge which Aristotle admitted had to be built up slowly through immediate apprehension or intuition;[16] and this was perhaps all that Plato meant by intelligible forms. There was therefore not a great difference between Aristotle's thought and that of his teacher, Plato. Both sought to accommodate a concept of universals divorced from the notion of corporeal being held by their predecessors. In a later terminology, both were realists in the sense that neither Ideas nor essences were mere concepts or even words but formed an ontological hierarchy. Aristotle's realism, however, was not unequivocal. In his preoccupation with predication and with objects of thought as distinct from substance designated as a *this*, he had sown the seeds for the major re-orientation of thought which was to come many centuries later.

Aristotelianism eventually became the philosophic orthodoxy in both the Christian and Islamic worlds and transmitted the influence also of Plato and through him of the Pythagoreans, Eleatics and Heraclitus. By the time it lost this prominence during the 16th and 17th centuries, it had influenced the whole tenor of western thought. Aristotle's syllogistic logic, for instance, though it has rivals, is still in use today and presupposes his ontological hierarchy of essences and genera.[17] In economics the influence of Aristotelianism and its antecedents is seen, for instance, in the widely held belief before the advent of neo-classical economics that political economy should properly concern itself with value rather than with the ever-shifting prices of the phenomenal world. It is evident, as we shall see, in Menger's idea of economics as an exact discipline and in the work of one of his intellectual successors, von Mises. It is seen in Marx's idea of forces of production operating in an ideal sphere and carrying with them periodically and cataclysmically changing superstructures in the phenomenal world of which the details can take on various forms.

Even the neo-classical economist who does not regard a model as merely one of an indefinite number of possible mental constructs to serve an ad hoc purpose, must think that his theories and models somehow capture essences that are also out there in the restless welter of everyday life. Some economists with a casual interest in methodology seem almost to take pride in denouncing essence as a meaningless notion. But it is quite likely that the very same economists speak about definition and abstraction without any misgivings and then, unless they are unusually imbued with the spirit of modern axiomatics, they are thinking in terms of what Aristotle called essences.

3. Knower and known

Among the schoolmen of medieval times there was an interminable debate between (in a later terminology) realists on the one hand and conceptualists and nominalists on the other. The issue was whether universals have a real existence or whether they are concepts or even mere words or names, i.e. whether they are discovered or invented.[18] Medieval scholasticism is often thought, probably unfairly, to have been characterized by an unbearable casuistry. But the issues in the realism–nominalism debate are significant even today because a species of nominalism in the form of positivism was prominent for much of this century. The nominalist holds that a universal is a mere word or name, with the implication that it is a mere fiction or at best an arbitrary mental construct and can be created at will. But since it is not his intention to say that we create our experience at will, the *given* or the ultimate arbiter of reality is sought in our *immediate sensory experience* or *sense data*. However, the question of what immediate sensory experience is devoid of mental constructs is no less perplexing than the question of what real essences are.

The evidence of the senses has of course always been the cornerstone of empiricism, but its significance has differed. In antiquity empiricism often led to scepticisms, to the conclusion that there is nothing to know because everything always changes. The empiricism that arose some 300 to 400 years ago did not incline to such negative conclusions. The great advances made in the physical sciences had had an immense impact on thought. The presumption now was that nature was orderly and regular and that the only problem was to understand how men had got to know this order.

To Galileo the key was mathematics, for nature was to him as it had been to the Pythagoreans essentially quantitative.[19] His view and that of some of his contemporaries developed into classical mechanics, though its definitive formulation was due to Newton. In this view the universe consists of particles of matter in motion, not in more or less disorderly commotion as in the ancient atomic theory, but in a motion governed by definite laws. Classical mechanics became the new orthodoxy and in a generalized form, as will be argued in later chapters, still exercises an influence over economics. The problem of similarity did not really exist in the mechanical universe because, apart from size and shape, there simply were no qualities. Furthermore, there was no place in it for minds, for purposes and for ethics. The obvious solution was to create another world for all these familiar things, a solution which is usually referred to as dualism. Minds were somehow outside the physical universe but became conscious of it in terms of similarities and differences in qualities which, however, were reducible to quantitative terms, rather as differences in colour are reducible to differences in wavelength or frequency.

Dualism assumed a different character in the hands of Descartes. There was more to knowledge than a mere deciphering of messages crossing the border between the material and mental worlds. Apodictic knowledge was gained from insight and reason, just as one's being was known from the indubitable fact of consciousness – the famous *cogito* principle, which in a way was reminiscent of the *esti* of Parmenides but also introduced a subjective element which was to change the whole tenor of western philosophy. Certain basic universals could be obtained from pure thought (i.e. without observation) and in terms of these – at least that was the ideal – the nature of reality could be demonstrated clearly by pure reasoning. The doctrine was developed further by Leibniz and reformulated a kind of Platonic realism in terms of innate ideas. It has come to be known as rationalism and still has many adherents, some even in economics.[20]

The notion of knowledge derived from innate ideas was unacceptable to empiricists. Locke's famous *Essay* was directed against the notion of innate ideas. The sources of ideas, according to Locke, were sensations and feelings. The picture he drew was based on the dualism conceived by Galileo. Particles of matter impinged on the senses and through the medium of nerves and the brain produced *simple ideas* in the mind. Corresponding to simple ideas in the mind there were qualities of *external objects*. A distinction had to be made between qualities such as size and shape which were compatible with the mechanistic conception and which he called primary qualities, and others such as colour, sound, temperature, taste which he called secondary qualities. Primary qualities were actually part of the external objects, but secondary qualities were appearances produced in the mind by 'the operation of insensible particles on our Senses'. Simple ideas were built up into compound or complex ones (such as objects) and these were grouped into *sorts* according to *nominal essences* which corresponded to man-made concepts or, which was the same to Locke, to words.[21]

Locke's secondary qualities were appearances by which the underlying physical reality of matter in motion manifested itself in minds. The idea had already been a feature of Galileo's thought and had, in fact, been expounded by Democritus in antiquity. To Berkeley it seemed to be the contrary of what common sense told one. Adopting a position philosophers label as idealist, Berkeley held that just as thoughts and feelings cannot exist without minds so external objects inferred from sensations cannot exist except as perceived by someone. This was his *New Principle*, that *esse* is *percipi*. He used the principle to argue against the presuppositions of classical mechanics. Far from being the underlying reality, a universe of unperceived inert matter in motion was a pernicious abstraction and a mere playing with words. 'For can there be a nicer strain of abstraction than to distinguish the existence of sensible objects from their being perceived, so as to conceive them existing unperceived?' To him the mechanistic conception was 'either a direct contradiction, or else nothing at all'.[22] The theological conclusions Berkeley drew from his analysis have been less influential, but the New Principle, according to which one did not try to reconcile mind with an independently existing nature but considered the relativity of the known to the knowing mind, has been very influential. In essentials it is what Kant was to call a Copernican revolution in thought. It promoted a subjective approach which was to influence both subsequent idealism and positivism.

This influence was already apparent in Hume's *Treatise*, a precursor of positivism. In contrast to Locke, Hume held that it was 'perfectly inexplicable by human reason' how mental impressions arose from the senses. Nor did it matter because inferences could be drawn from perceptions irrespective of 'whether they represent nature justly'.[23] This is almost identical with the way the contemporary positivist, A.J. Ayer, formulated his *sense-datum language* some years ago.[24] Hume's most characteristic approach was to consider the association of ideas. His justly most celebrated conclusion from this was the problem of induction, that causation or the notion of necessary connection was not based on reason but on a habit of thought formed when constant conjunctions (of impressions or ideas) were noticed. In the 'Abstract' of the *Treatise* he illustrated the point with an example of billiard balls imparting motion on impact – the significance of the idea for classical mechanics was therefore not missed. The 'Abstract' ends with a commendation of the principles of the association of ideas, namely, resemblance, contiguity and causation: 'so far as regards the mind, these are the only links that bind the parts of the universe together or connect us with any person or object exterior to ourselves ... they are really to us the cement of the universe'.[25] From there it was a small step to the *mental expedients* and *mental economy* which Mach invoked towards the end of the 19th century in his positivist critique of Newton's concepts and therefore of classical mechanics.[26]

However, the more the metaphysical underpinning of empiricism was turned into a mental construct, the more difficult it became to say what constituted the *given* on which all knowledge was to be based. It is here that the problem of similarity reappears. Locke had observed that, though all things that exist are particular things, languages do not consist of proper names. Nor could they; to 'heap up names of particular things' would not allow communication at all. Words were signs for general ideas and these are formed quite arbitrarily by combining the common qualities of various things.[27] But, as Leibniz was quick to point out, to allow for common qualities was to bring realism in by the back door, even though it was Locke's intention to remove real essences from the *given*.[28] Berkeley had equated concepts with images and contended that it was not possible to have a concept (image) of, say, a general man but only of a particular man. Abstract ideas were therefore particular ideas which represented other particular ideas just as a single word could stand for varied things.[29] But concepts are not necessarily images and *representing* also brings real essences in by the back door.

One way out of these difficulties was indicated by the philosophical reconciliation proposed by Kant after he had met Hume. The *cement of the universe* of Hume, being a mere habit of thought, could not be the basis of the necessary or certain propositions that mathematics and at one time the Newtonian scheme were believed to provide. But Kant would not accept Leibniz's rationalism uncritically either. He was impressed by the fact that there seemed to be certain forms of thought or categories (e.g. space, time, causation, etc.) which persisted throughout the constantly changing content of experience. The categories, he held, were the mind's own contribution to knowledge – they so to say acted upon unknowable *things in themselves* to produce the phenomena of experience. Categories were not found in phenomena by the senses but put into them by the mind and, for this reason, he believed, synthetic a priori propositions were possible, i.e. propositions that are true even though they are neither tautological (synthetic) nor based on observation (a priori).[30] The categories Kant listed do not cover all cases in which the problem of similarity arises. He did, however, make some remarks to the effect that the universals according to which things are classified could also be intuitive forms of thought.[31] This suggestion was taken up by some of the many neo-Kantians who were influential in the latter part of the 19th century and the early years of the 20th. In the formulation of the Marburg School, facts are never *given* but *taken* or *arrived at* in the sense that they depend on the questions set by the mind, which in turn depend on the a priori concepts used.

The positivists were not prepared to accept this blend of idealism and realism. Nevertheless, their approach was similar to that of the neo-Kantians (with whom, however, they were often at loggerheads) in so far as they also held that the mind contributed something to knowledge. But it contributed only words or more or less arbitrary constructs which, when shared by many, became conceptual conventions or languages used for organizing and ordering facts. The physicist, Ernst Mach, adopted a thoroughgoing nominalist position. Science accomplished an ordering of sensations to achieve certain purposes and good science strove for simplicity and economy in the use of concepts – an idea that had been propounded, as is well known, by the medieval nominalist, William of Ockham. On this basis, Mach criticized the Newtonian conception of absolute time and space and even the belief in the real existence of atoms.[32] He and like-minded scientists were eager to show that scientific theorizing was a creative activity and, in this endeavour, bare facts or immediate experience tended to be taken for granted. In explaining the need for an imaginative selection of facts, Henri Poincare remarked that 'while the scientist is discovering one fact, millions and millions are produced in every cubic inch of his body'.[33] It is not clear how facts are here conceived, but one gathers that they are not only in us but all around us, like grains of sand to someone on a beach. The need for selection and ordering is then obvious.

But this clear-cut distinction between creations of the mind and given facts cannot always be maintained. There is a double requirement to be met. Since theories are invented but knowledge is not, there must be discovered facts, i.e. theory-free facts. On the other hand, facts must somehow be fitted into theories so that they may be ordered, and implications may be drawn from them. What this double requirement involves may be seen more clearly when, in the style of the 20th-century philosophy, the matter is expressed in terms of statements. There are synthetic and analytic statements. Synthetic statements designate facts, analytic statements are tautologies, i.e. statements in which the subject logically entails the predicate, and it is these that human ingenuity creates in devising theoretical frameworks for ordering facts. For example, a positivist regards micro-economic theory as a set of interrelated analytic statements. However, since all statements are predications, they all assert or deny that an attribute or quality attaches to something. Even synthetic statements do not express a *this* but a *such*, i.e. they are framed in terms of universals (as Aristotle pointed out so long ago). The double requirement mentioned above therefore boils down to the one requirement that the universals in synthetic and analytic statements be the same. If they are not, theories and facts are simply unconnected.[34] If they are, both types of statement may be made in the same language; facts may be ordered and statements that purport to be about something may be tested by a recourse to facts. But it is not easy to explain why the given, which by the very meaning of the term is beyond our control, should fit in so neatly with our conceptual creations. The supposition that concepts are inherent in the given leads to the undesired conclusion that our apparent freedom to create (empirical) concepts is an illusion and that we must after all confine ourselves to discovering the Aristotelian hierarchy of real essences. If concepts are not inherent in the given, one is left with the problem of how some ultimate, theory-free, conceptless given can be translated into statements with constructed concepts.

Positivists have been rather vague about this problem. To Mach, similarity of sensations was based on what appeared to be the neurological possibilities, so that for similar sensations there corresponded similar *nerve processes*. Concepts were somehow built up from similar sensations, yet they were not so tied to them that sensations deny us the freedom to choose the best means for achieving conceptual economy.[35] In his chief work, Moritz Schlick, who was to become the leading member of the *Vienna Circle* of logical positivists, rejected Mach's analysis of sensations. But in an otherwise closely reasoned and sophisticated book, he was nevertheless vague about the given. 'Everything that is given to us from the world is given to us in intuition.' But in intuition things 'are only given, not understood'. To be known and understood, they must be 'incorporated in some way into a conceptual system'. This entails a 'process of comparing, finding again and designating'. Designation is 'a quite arbitrary stipulation' that a sign corresponds to something given and the sign no more pictures the given than musical notation pictures tones. Designation serves a purpose only if we can 'find again'. At the beginning of the book Schlick said he would leave the question of how we manage to see similarities to psychologists. However, he ended the book by remarking that knowledge 'would not be possible if there were no sameness's' and concluded that it must be that reality is just so constituted.[36]

Nor can much on this matter be gleaned from the early Wittgenstein's *Tractatus*,[37] which strongly influenced the logical positivists. His interest really lay in the uses and limits of logic, but the following standpoint can be pieced together. 'The world is the totality of facts, not of things.' 'We picture facts to ourselves.' 'A picture can depict any reality whose form it has.' 'In order to tell whether a picture is true or false we must compare it with reality.' Or again, propositions are pictures of reality, i.e. models of reality as we imagine it. They are truth-functions of elementary propositions which cannot be derived from each other (e.g. predicted) and cannot contradict each other. An elementary proposition asserts a state of affairs. If it is true the state of affairs exists, otherwise not.[38] Wittgenstein held that we know on logical grounds that analysis of propositions necessarily leads to elementary propositions, but he did not venture any examples of such propositions. Richard von Mises, however, did venture to illustrate the irreducible expression of the given. It 'consists only of very short, very simple expressions corresponding to immediate perceptions, e.g. "here pointer at five", "there red spot of short duration", etc. But even a word like *pointer* already says too much. There should really be only a denotation of a certain movable something'. He conceded that there were differences of opinion about how the given could be stated. But these 'do not touch the core of the matter and can be resolved satisfactorily'.[39]

In its further development, especially in the work of Carnap, logical positivism became mainly an investigation of syntax, of the logical rules for translating one kind of statement into another kind, and less emphasis was put on the given. In Carnap's early work the given was already regarded as something artificially separated from experience.[40] In his later work, the criterion for meaningfulness became conceptual coherence rather than reduction to the given, which in many cases he held to be impossible. The problem, as he and Neurath saw it, was that certain statements, e.g. those in mathematics and many in physics, are only possible on the basis of *unlimited operators* or pure postulates which are linked to descriptions of experience only indirectly because they logically exclude certain possibilities of experience. Statements with such postulates are therefore incompletely testable in the sense that they can only be falsified, but Carnap held that in a case of falsification we are still left the choice either of rejecting the postulate or of rejecting the description of the given. To test a hypothesis is then to investigate whether a statement under consideration (i.e. the hypothesis) is logically consistent with other statements we make.[41] In this he entered the sphere of pragmatism, since the criterion for validity can then only be something like William James' dictum that a hypothesis is true if it is useful and useful if it is true.[42] Not all positivists, however, were prepared to follow him on this road.[43]

Popper (who does not regard himself as a positivist) put forward a scheme similar to Carnap's. Hypotheses may be framed in terms of universal statements which logically exclude certain particular occurrences. Hypotheses are falsified if the occurrences they exclude are verified and a falsifying hypothesis is corroborated. Verification is thus not eliminated and the problem of how elementary propositions (Popper calls them basic or test statements) are derived from some kind of reality not under our control remains unresolved. Popper the logician, however, was not interested in *psychologistic* (subjective) questions about individual perception, but only in the logical properties of test statements and of universal statements that are scientific, and in the logical relations between such statements. Test statements must simply be assumed to exist.[44] In a recent restatement of his position, Popper says that he will not consider the question: 'How do we decide the truth or falsity of test statements?' Elsewhere in the same book, however, he does say that 'there can be no pure observational language' and 'there are no such things as sense data or perceptions which are not built upon theories'. 'Sense organs incorporate the equivalent of primitive and uncritically accepted theories, which are less widely tested than scientific theories.' But if that is the case, one may wonder by what criteria either scientific or primitive theories can possibly be tested. Popper says that 'the assumption of the truth of test statements sometimes allows us to justify the claim that an explanatory theory is false'. One may well feel that his principle of demarcation between science and metaphysics and his claims to have solved the problem of induction for mankind must surely suffer if falsification is based on assumption. Popper does admit that 'we have no criterion of truth, and no means of being even quite sure of the falsity of a theory'. But none of this is very important to Popper anymore. In the *objective* and *evolutionary* approach he has adopted, it does not really matter what we are sure of or what we think are data, because 'the alleged data are in fact adaptive reactions' (of organisms), i.e. the interaction of the environment with the theory-equivalents in our sense organs in which rough edges left by natural selection are rubbed off. On the other hand, Popper also teaches that all explanatory theories are conjectures, mere guesses, and this presumably also applies to the evolutionary explanation.[45]

4. Positivism in economics

The falsifiability criterion has of course become well known among economists and has become identified with a kind of positivism grafted on a basically Aristotelian conception. Milton Friedman's essay on positive economics is probably the best-known statement of this kind of positivism.[46] The positivistic superstructure is familiar enough. Theory consists of analytic and synthetic statements. 'Viewed as a language, theory … is a set of tautologies. Its function is to serve as a filing system for organizing empirical material' (p 7). 'Viewed as a body of substantive hypotheses, theory is to be judged by its predictive power' (p 8). 'If there is one hypothesis that is consistent with the available evidence, there are always an infinite number that are' (p 9). Choice among hypotheses is to be made on the basis of 'economy, clarity and precision' (p 40).

Underlying all this, however, there is a recognition of real essences. 'Economics as a positive science is a body of tentatively accepted generalisations about economic phenomena' (p 39). 'A hypothesis is important … if it abstracts the common and crucial elements from the mass of complex and detailed circumstances surrounding the phenomena to be explained' (p 14). 'Factual evidence alone can show whether the categories of the *analytical filing system* have a meaningful empirical counterpart' (p 7;

original emphasis). 'There is seldom much doubt whether a particular factor should be classified as affecting supply … or demand' (p 8; in other words, particular factors have intuitively obvious common qualities). None of this could make sense to someone who simply did not believe in essences and the attendant species and genera.

It soon appears, however, that the efficient filing system and the hierarchy of essences are not meant to coincide, as one of the above remarks could suggest. Rather, positivism seems to be reserved for drawing up hypotheses and essences for testing them. This comes out in the discussion on the *realism* of assumptions (i.e. broadly, whether assumptions are real essences). Friedman believes that an important hypothesis 'must be descriptively false in its assumptions' (p 14). At times he suggests that this must be so because assumptions are abstractions. But this does not always do. The real question is whether assumptions 'are sufficiently good approximations for the purpose in hand' which 'can be answered only by seeing whether the theory works, which means whether it yields sufficiently accurate predictions' (p 15). In fact, he also argues that 'to suppose that hypotheses have not only *implications* but also *assumptions* … is fundamentally wrong and productive of much mischief' (p 14; original emphasis). In other words, the assumptions of a hypothesis are nothing but the hypothesis itself and the only way to judge the *realism* of the assumptions/hypothesis is to see whether the implications remain unfalsified. He gives some examples. The density of leaves on a tree can conveniently be explained by saying that leaves assume positions *as if* they sought to maximize the amount of sunlight they receive; billiard shots by saying that billiard players exercise their skills as if they did complicated mathematical calculations. Likewise, the actions of entrepreneurs can be explained as if they equated marginal cost with marginal revenue and many economic questions as if there were perfect competition.

Friedman's non-realistic assumptions (and to him most of the postulates of neo-classical economic theory are apparently such) are similar to the unlimited operators or general postulates of the later logical positivists or to theories in Popper's sense. They cannot be formed by abstraction because they are *not in* descriptions of experience. But they may be testable because they may have implications (predictions) that are amenable to observation. At this point one seems to return to Aristotle and the presupposition that observation yields facts which are readily identifiable by their nature, i.e. which appear in the form of real essences and can therefore easily be classified and presented as statistics. Thus, economists of this persuasion can test hypotheses about the allocation of time, human capital or whatever,[47] secure in the knowledge that assumptions do not have to be realistic (real essences) but that happily one can always frame theories in such a way that they have implications among directly intuited real essences. In many cases, in fact, one does not have to bother about Popper-style falsification at all. Hypotheses relating income to education or price level to money supply can be tested (so it would seem from the work of these economists) in the manner of Aristotelian logic, i.e. on the understanding that a general hypothesis has particular instances and that it is falsified if a particular instance fails to occur.

Positivism in economics is therefore a blend of nominalism and realism. Whether it is a blend which offers the best of both worlds is arguable. One reason for doubt is (to extend the metaphor) that it is a blend without fixed proportions, that everyone is his own blender. The point should be elaborated. The testing of hypotheses would seem to require, apart from the logical relations which interested Popper, that the tester sees a difference between the contrived and the discovered, or between the filing system and the things to be filed. But how clearly are such differences seen? After all, we can argue about what should be included in money supply or what constitutes education and income. Can we appeal to essential natures in settling such arguments, or do we only have conventions to be judged by their usefulness as tools for prediction? Income as a concept is said to be of comparatively recent origin, there having been nothing corresponding to it in the ancient and medieval worlds. If this is so, then was there always something which in essence was income, though no one realized it, or did the income concept arise out of a certain analysis of economic affairs as a classificatory expedient defined more or less by a specification of types of items? If the latter is the case, the essence-versus-expedient question must be posed for the types of items classified (receipt from work, business, assets, etc., or from the creation of utility) for we can agree that the notion of testing can make sense only if we eventually encounter something that is actually a datum rather than a mental expedient. It may seem sensible to say that in this regard each tester may be left to exercise his judgement and draw the line between the contrived and the real where he thinks it ought to be, or perhaps where at the time it suits him best that it should be. But such freedom of choice must surely invite the criticism that this is a positivism without conviction, that it is merely a licence for unfettered conjecture when the apparently real becomes frustratingly perplexing, or even, that it is an open invitation for cooking the books.

Friedman himself seems to have had something else in mind. In countering the argument that economic phenomena are necessarily varied, and complex, he remarked: 'A theory is the way we perceive *facts*, and we cannot perceive *facts* without a theory' (p 34; original emphasis). But this remark, sensible as it may be, does not help the positivistic case. We have seen that according to Friedman, theory may be viewed either as a language serving the function of a filing system or as 'a body of substantive hypotheses'. If, therefore, a fact becomes a fact only on being filed, on what basis is it put into one file rather than another? If, on the other hand, facts presuppose *substantive hypotheses*, against what are the latter ultimately tested? It seems that one has somehow got oneself locked into the filing cabinet.

Unfortunately, the way of the purist in positivism does not seem to be any more promising. To such a purist there are presumably no essences of income, education, money, etc.; these terms refer to constructions we put on things. But on what? A sensible person might suppose that it is on what people are trying to do and on what happens to them owing to the action of others. But can we perceive aims or purposes directly? That would come close to accepting Socratic forms or Platonic Ideas. Surely, we merely put constructions on what people are observed to do. But can we speak of people? The human essence is surely also a construct. One may wish to stop this process when one comes to descriptions of elementary sensations. However, one may be left wondering by what feats of mental ordering even Richard von Mises could have used a 'there soft, warm, moving something' to test hypotheses about income, education and money.

In a recent interview, Sir Alfred Ayer was asked what in retrospect were the major shortcomings of positivism. He replied that its 'defects were that nearly all of it was false' but 'that it was true in spirit, that the attitude was right', that 'what survives is the general rightness of the approach'.[48] It is the approach that has been most influential and, in so far as it has discouraged pure conjecture and wild speculation, it has no doubt been a salutary influence. In economics, unfortunately, its influence seems by now to be very nearly the opposite. The majority of economists, one imagines, feel no allegiance to positivism, pure or Friedman-type, nor to the philosophy of Sir Karl Popper. Yet many have been encouraged, it seems, to regard their subject as a conceptual tool-box full of models that one tries to fit to facts, rather as a mechanic may try out a number of spanners to see which one fits. It is ironic that a philosophical movement, motivated from the start by an opposition to metaphysical speculation, should have had the effect in economics of encouraging a proliferation of speculative models justified only by a vague notion that obliging econometricians, presumably with direct access to theory-free data, will someday put them all to the test.

5. Commitment, paradigms and knowing how

'The unit of empirical significance is the whole of science.' This sentence appears in a thought-provoking paper by Quine (1953, p 42) and is characteristic of some interesting developments in the theory of knowledge at the time. To Hume the units of empirical significance had been ideas or terms which had to be matched one to one by sense impressions. After Frege, the proposition or statement rather than the individual terms in it became the unit of analysis and the irreducible form of the given. Quine (1953, p 42), however, suggested that we have still 'drawn our grid too finely'.

Not the single proposition but the whole body of propositions making up a science should be related to experience. All knowledge 'from the most casual matters of geography and history to the profoundest laws of atomic physics or even of pure mathematics and logic, is a man-made fabric which impinges on experience only along the edges'.

The significance of this is that the relation between experience and knowledge is seen to be not as simple as doctrines of one-to-one correspondence make out, because we have 'much latitude of choice as to what statements to re-evaluate in the light of any single contrary experience'. The mention of logic and mathematics, apparently on the same footing as physics and even history, shows that Quine thought that even these were not immune to revision should we be unwilling to make adjustments elsewhere. (Revision of standard logic has apparently been proposed in the field of quantum mechanics.) In fact, it was one of Quine's major contentions (as it had been of some others) that the distinction made by positivists between analytic and synthetic statements, i.e. between logical and factual truths, is ultimately without foundation. 'Any statement can be held true come what may, if we make drastic enough adjustments elsewhere in the system.' It is 'our natural tendency to disturb the total system as little as possible' that leads us to believe that only some of our statements are liable to revision in the light of experience.[49]

These surprising conclusions were already latent in pragmatism and in a doctrine known as conventionalism, which goes back to Poincare and Duhem and is yet another offshoot from the philosophy of Kant. There is an example that has often been used to illustrate the central point of conventionalism.[50] Suppose we have heated different lots of a substance we understand to be phosphorus and have made the empirical generalization that the melting-point of phosphorus is 44°C. One day, however, we discover that a particular piece of phosphorus does not melt at 44°C. Do we simply drop our generalization? Initially at least this is very unlikely because there are many ways in which we can *save* the melting-point hypothesis. We may, for instance, suppose that the substance in question only appears to be phosphorus but is really something else. Or we may posit the presence of some other factor that has changed the melting-point; perhaps a catalyst of some kind or some external condition analogous to air pressure in the case of the boiling-point water. Starting off by perhaps questioning the accuracy of the measuring instruments, we may well use different stratagems for saving the hypothesis as the matter is investigated further. But as long as we go on saving the hypothesis, or, in the extreme case, as long as we will not accept a substance as phosphorus unless its melting point is 44°C (i.e. the melting-point is a necessary condition for identifying phosphorus), the melting-point hypothesis is not an empirical or synthetic proposition at all. It has been made an analytical proposition that cannot possibly be false.

Whether and for how long we regard a hypothesis as analytical or synthetic is then seen to be at least to some extent a matter of temperament and of our own decision.

There is much scope for conventionalist stratagems in economics. Suppose we wanted to test hypotheses about liquidity preference and interest rates and found that during a particular period interest rates actually fell while money supply decreased. We may then suppose that statistical coverage had been incomplete or the definition (identification) of money supply inadequate. Perhaps we would suspect a drastic fall in incomes or posit the existence of institutional factors that account for the anomaly. Only if, in analysing subsequent periods, we began to feel that such explanations were somehow unsatisfactory, would we suspect the hypotheses themselves. Only in a very extreme case would anyone even think of questioning the mathematical procedures and the logic involved (supposing them to have been correctly applied), for that in Quine's way of thinking would be to disturb the system violently indeed.

Suppose again that we came across a man who, having been offered two professional appointments, chose the lower paid of the two. Would we revise all economic propositions based on the hypothesis that the supply of labour is an increasing function of rates of remuneration? Our 'natural tendency to disturb the total system as little as possible' would be likely to come to the fore and we would say that the chosen appointment carried with it certain non-pecuniary benefits, i.e. that there was something the man particularly liked about the job he chose. We may then be satisfied that the observed facts did not contradict our hypothesis at all, though it might have been more accurate to say that we observed the facts we did because we were not prepared to countenance a contradiction of our hypothesis. The point of all this is that it may be idle to look for a simple juxtaposition of givens and mental constructs, that the positivist's injunction that we should hold our beliefs up to the facts may not be very useful and that *the facts* are themselves intimately bound up with the whole corpus of our beliefs or with our conceptions. This does not of course solve the problem of what is the given; indeed, it makes the problem more intractable.

When we identify things or facts, we are not normally conscious of applying any definite criteria, and in those cases in which we are so conscious we really need further criteria for identifying the criteria and so on until eventually we have to make do without them. In an Aristotelian manner of speaking, one could say that we have to rely on a direct apprehension of essences. But when this matter was reconsidered in the light of what we have just seen, i.e. that our perception of things and facts is tied up with our own preconceptions, some reformulation seemed to be called for. In this regard, a felicitous distinction and accompanying phraseology introduced by Ryle comes in useful. Ryle distinguished between knowing how and knowing that, or between being able to perform a task and having knowledge formulated in statements.[51] The ability to perform a task does not necessarily imply a knowledge of the rules and principles involved (e.g. the ability to speak a language, or to reason). Ryle wanted to equate intelligence with knowing how. But though Ryle did not do so, one could also apply the distinction to the problem of similarity. Knowing that presupposes recognition or the discernment of similarity. At some stage, therefore, recognition cannot be based on statements setting out criteria for identification (a knowing that). But it may be regarded as an ability. We may know how to recognize similar things without being able to state in which respects they are similar. For example, we would be most surprised to find a person who did not know how to tell a dog from a cat or a cow. Yet, if asked what it is that all dogs have in common but that no other animal has, how many people could find an answer? Zoologists no doubt have criteria, but it is doubtful whether even a zoologist walking down the street would recognize a dog by them. One may say that we simply have a disposition or a propensity to discern similarities and to identify. This does not of course solve the problem of similarity, but it puts it beyond the scope of logical analysis.

In recent years much has been made of dispositions to learn or to formulate knowledge. The idea, however, goes back to Hume and there were hints of it in the doctrine of innate ideas of Leibniz and Descartes.[52] The linguist and philosopher, Noam Chomsky, has recently revived this rationalist notion of inborn disposition in his study of common forms in natural languages. Dispositions are therefore open to interpretations in terms of genetics, and some have concentrated on this aspect. Popper, for instance, deviating a little from his earlier position, has ventured the estimate that 99.9 per cent of knowledge (in the form of universal statements) is transmitted genetically.[53]

Michael Polanyi, the physical chemist turned philosopher, emphasized a quite different aspect of dispositions. He was opposed to what he saw as an ideal of scientific detachment and as the objectivity of an impersonal knowledge. In its stead he proposed an ideal of personal knowledge which stressed 'the personal participation of the knower in all acts of understanding'. This participation manifests itself both in the *intellectual passion* for order (finding expression in rationalism, whose Platonic and even Pythagorean origins he acknowledged) and in the human terms in which knowledge is formulated. 'To a disembodied intellect, entirely incapable of lust, pain or comfort, most of our vocabulary would be incomprehensible.' Polanyi was impressed by studies of intelligence in animals. They led him to the conclusion that man's intellectual superiority lay only in his ability to articulate thought into concepts, symbols and language. Articulation 'disciplines and expands the reasoning powers of man'. It turns a knowing how into a knowing that, or, as he put it, a tacit component of knowledge into a formal one. But the formal component is merely a superstructure built on to the tacit component, i.e. on to the disposition or ability to perform mental tasks, which is also found in animals. In a sense, words in themselves have no meaning, but speakers and listener mean something by them – they practice *the art of knowing*. Conceptions act as self-set standards by which experience is articulated into new instances of familiar entities.

Polanyi held that the attempt to verify or falsify statements objectively leads to an infinite regress because it tries to make the assertion that a statement is true or consistent with the facts into yet another factual statement requiring further verification or testing. This is avoided if it is accepted that such an assertion is not a statement but a personal commitment, a claim to have established contact with reality. Polanyi was at pains to explain himself. Commitment is not a matter of whim or fancy but of conviction. It is a personal act with *universal intent*. He illustrated this by analogy. When a judge has to decide a case not covered by legal precedent or statute, his freedom to do as he pleases is nevertheless restricted by his commitment to the ideal of justice. Likewise, comprehension 'is neither an arbitrary act nor a passive experience, but a responsible act claiming universal validity'. The spirit of commitment was expressed perfectly, Polanyi thought, by Luther's famous profession: 'Here I stand and cannot otherwise.'

As an activity, commitment is necessarily fallible. 'Only an activity can go wrong, and all activity incurs the risk of failure.' Nevertheless, a sincerely held set of beliefs has a certain stability because references to reality are constructed from it. Polanyi called this the circularity of a conceptual framework. He dwelt at some length on the Evans-Pritchard study of the Azande (a seemingly never-ending source of inspiration for social thought) which described how the members of this primitive tribe derived all their explanations of events from their belief in witchcraft and how they maintained their belief by what were referred to above as conventionalist stratagems. With suitable ad hoc explanations, witchcraft cannot be faulted. To Polanyi this illustrated how a sincerely held belief both shapes the understanding and reinforces itself. He then quoted from the writings of a former Marxist and a former Freudian, both of whom, reflecting on their former intellectual allegiances, seemed to make the point that 'the all-embracing interpretative powers' of a conceptual framework is taken 'as evidence of its truth'. With his background in chemistry, Polanyi could give examples of circularity in this field as well.[54]

Thomas S. Kuhn acknowledged the similarity between his normal science based on a paradigm and Polanyi's tacit knowledge. However, the interest of a historian of science whose line of thought is similar to Polanyi's, is naturally centred in the circumstances in which the stability of beliefs breaks down and new beliefs are adopted. It is to be expected that in this context more stress would be laid on acquired rather than on innate dispositions. Normal science is a cultural phenomenon. It is intersubjective and acquired through social contact. Kuhn's main thesis is not of direct interest here, but the ideas of a paradigm and of normal science which emerge from it should be considered briefly. Kuhn originally referred to an influential work such as Newton's *Principia* as a paradigm because it inspired others to use it as an example and thereby to create an orthodoxy or normal science. However, the word *paradigm*, as he later admitted, was used rather loosely. Kuhn tried to remedy this in his *Postscript – 1969*.[55]

The term *paradigm*, he says there, was used in a wider and a narrower sense in his original text, though it should properly have been used only in the latter. For the wider sense he now proposes the term *disciplinary matrix*, though 'constellation of group commitments' also appears. This includes formal *symbolic generalizations*, a shared belief in certain models, shading off into a belief in the aptness of certain metaphors, and shared values such as that theories must be able to predict or that they either must be or need not be *socially useful*. For the narrower sense, which he regards as the more important and which first led him to the word *paradigm*, he now proposes the term *exemplar*. 'The practice of normal science depends on the ability, acquired from exemplars, to group objects and situations into similarity sets which are primitive in the sense that the grouping is done without an answer to the question, "Similar with respect to what?"'

There are no criteria for identification (i.e. for grouping into similarity sets) because paradigms have roots in an unarticulated knowledge of how to practice a discipline and this knowledge is acquired by doing rather than by learning rules. In training one learns 'from problems to see situations as like each other'. After the initial training has been completed, one has 'assimilated a time-tested and group-licensed way of seeing'. Thus, objects and situations are not complete entities, seen by everyone alike, which are simply grouped differently according to different paradigms. To see new *similarity relations* is to see new entities and what one sees depends on the paradigm one is following. 'Individuals raised in different societies' sometimes behave as though they see different things while, so to say, looking at the same thing. 'If we were not tempted to identify stimuli one-to-one with sensations, we might recognize that they actually do so.' But Kuhn is adamant that this does not imply relativism, that all paradigms are equally valid. 'In many environments a group that could not tell wolves from dogs could not endure. … It is just because so very few ways of seeing will do that the ones that have withstood the tests of group use are worth transmitting from generation to generation.'

6. Epilogue

Here we shall leave the development of the theory of knowledge. The ideas which have been reviewed will be used both to identify the apparent presuppositions of the creators of micro-economic theories, for economists also have worked in the milieu of western thought, and to consider the question of the empirical content of economics. While a selection such as this is bound to reflect the views not only of the selected but also of the selector, it has been presented in the belief that it has traced out at least a major stream of epistemological thought.

We have seen how some 2,500 years ago Greek thinkers grappled with the profound problem of being and becoming in terms of a conception of animistic, corporeal existence no doubt inherited from mythological cosmogonies. Out of this came the apprehension of an intelligible, logical realm which was to split the knowable into intelligible and phenomenal compartments. In the synthesis of Aristotle this apprehension gave just a hint of that re-orientation of thought which has been termed subjectivism and whose working out has been perhaps the major contribution of the last 300 to 400 years. This re-orientation reached a high point in the philosophy of Kant, but nevertheless continued thereafter to be a gradual evolution, sharing men's allegiance with the generalizations of classical mechanics. Positivism may be seen as a doctrine characteristic of the transition. It tried to anchor the intelligible in what was left of that conception of an external world which at one time had contained all that there was. This led to difficulties that might have been foreseen had more heed been paid to Aristotle's analysis. Of greater consequence, however, was the tendency for positivism to make of the intelligible something arbitrary and subject to whim, though mathematics and logic were given special exemptions. In recent years, the positivistic enterprise has been largely abandoned, at least in certain circles. The real essence of old has become the immediate perception according to innate dispositions, commitments to beliefs or intersubjective conceptions based on paradigms. In this, the re-orientation seems to be reaching at least one of its logical conclusions. The distinction between the intelligible and the phenomenal has once more become blurred, as it was in the days of Pythagoras and Heraclitus but to the latter this was so because everything had to have a place in the spatially and temporally extended universe whereas it is so now because everything can be known to us only through our human understanding.

Of course, by no means all who are concerned about such matters have accepted the subjective re-orientation. Popper has called it the *subjectivist blunder* of western philosophy. He praised Winston Churchill for having come up with 'the philosophically soundest and most ingenious argument against subjectivist epistemology that I know'. The argument is that an eclipse of the sun (according to Churchill) could be predicted by an *automatic calculating machine*, indicating apparently that the whole thing takes place independently of man's conceptualizing. Churchill therefore felt confident to reaffirm emphatically 'that the sun is real and also that it is hot – in fact as hot as Hell'.

Popper himself, as well as others, also seems on occasion to have argued that those who hold that perception (as distinct from theorizing) requires a conceptual medium, make the world into a kind of unreal dream.[56] If we are to understand *real* and *mental* in the same way as we understand *natural* and *man-made* as mutually exclusive opposites, then such conclusions seem indeed to follow. The apparent implication of calling conceptually mediated perceptions the real are expressed in those baser jibes about mountains that disappear whenever no one is looking at them. Yet the criticized no doubt are left with the feeling that their critics have misunderstood them. It is hard to see what more one can do than to try to distinguish between indulging in fantasies and adhering to criteria of objectivity. If that is not enough to earn for one's conclusions the description *objective*, because the conceptual medium and the personal assessment have not been removed, then it would seem that the term *objective knowledge* is simply self-contradictory.

Notes

1 Knight (1940).

2 Knight (1940, p 10).

3 Knight (1940, p 11).

4 The phraseology of the last two sentences has been adapted from that of Professor Ernest Gellner. See Gellner (1974, especially p 27).

5 The relativist would hold, e.g., that economists who respectively live in, or at least expound, Marxian, neo-classical, Keynesian, Austrian and many other worlds, are doomed forever to speak at cross purposes (when they do speak to each other). Gellner (1974, p 20), who is opposed to relativism, charges the late Wittgenstein with having encouraged relativistic notions in recent decades. Wittgenstein spoke of forms of life (*Lebensformen*) as the given (*das hinzunehmende Gegebene*). Relativism, however, has a long history. It was apparently taught by the sophist Protagoras in the 5th century BC.

6 See Kirk (1954, pp 69–70 and 74–75); Cherniss (1955, pp 35–46 and 352–372); Guthrie (1962, Vol I, pp 5, 62–64, 82, 89, 116 and elsewhere). The restricted sense of existence may perhaps be compared with restricted senses of production in the history of economic thought. Mercantilist writers regarded only export industries as productive. The Physiocrats, as is well known, equated product with tangible commodities and hence regarded only agriculture as productive. Adam Smith, though he criticized the Physiocrats for not recognizing manufacturing as productive, nevertheless restricted the meaning of production to the creation of vendible commodities. Servants, lawyers, judges, churchmen, physicians, musicians, etc. he called unproductive labour (Smith 1937 [1776], Book II, Chapter III).

7 Popper (1945, Vol I, pp 9–14).

8 Kirk (1954, pp 367–380); Guthrie (1962, Vol I, pp 489–492).

9 Heraclitus (or Herakleitos) of Ephesus is thought to have lived from about 540 to 480 BC. Only fragments of his book, gathered from other writers of antiquity, are extant. The remarks in the text are based mainly on the following works: Kirk (1954); Guthrie (1962, Vol I, pp 405–492); Kirk and Raven (1957); Freeman (1953). The probable meaning of the words *Logos* and *Kosmos* in the time of Heraclitus are discussed in Kirk (1954, pp 37–46, 67–71 and 307–316) and in Guthrie (1962, Vol I, pp 110, 208, 419–424 and 455).

10 Guthrie (1962, Vol II, pp 1–118); Kirk and Raven (1957, pp 263–285); Freeman (1953, pp 140–152).

11 Guthrie (1962, Vol II, pp 35–34 and 389–393).

12 Guthrie (1962, Vol I, pp 146–340).

13 Plato, *The Republic*, Book VII.

14 Since the extant works of Aristotle were compiled and edited about 260 years after his death, from short lecture notes written at various times throughout his life, they are somewhat repetitive and lack a clear expository progression. The following is therefore only a selection of the passages in which the matters discussed in the text are dealt with. On his critique of Plato's Ideas, see *Metaphysics* I 6 and 9, XIII 4–6; on his misgivings about the perceptible, *Posterior Analytics* I 33, *Metaphysics* IV 5; on substance, *Categories* 5 and *Metaphysics* VII 1–17; on knowledge always being of universals, *Posterior Analytics* I 31 and *Metaphysics* I 4 and 6; on essences and accidental attributes, *Posterior Analytics* I 4 and *Metaphysics* VII 3–6, VIII 1 and 6; on the four causes, *Physics* II 3, *Generation and Corruption* II 9 and *Metaphysics* V 2; on causes and essences, *Posterior Analytics* 2 and 8, *Metaphysics* VII 17.

15 The issue of the immanence versus the transcendence of Ideas is discussed in Collingwood (1945, pp 55–72 and 85–86).

16 *Posterior Analytics* II 19. No bibliographic reference.

17 See Cassirer (1953 [1910], pp 4–9 and 10–26 *passim*). See also Cohen and Nagel (1934, p v): 'We do not believe that there is any non-Aristotelian logic in the sense in which there is a non-Euclidian geometry ... alternative systems of logic are different system of notation or symbolization.'

18 A brief discussion of the debate may be found in Aaron (1967, pp 1–17).

19 See for instance the quotation from Galileo in Collingwood (1945, p 102).

20 For example, Hollis and Nell (1975) advocate a return to the rationalism of what they call the Classical-Marxian tradition in political economy. See especially Chapter 9, pp 233–266.

21 Locke (1975 [1690], p 136).

22 George Berkeley, *A Treatise concerning the Principles of Human Knowledge* (first published 1710). The quotations are from Berkeley (1901, Vol I, pp 258–260 and 270).

23 Hume (1962 [1739], Book I, Part III, Section V).

24 Ayer (1940). Almost the whole book concerns this question but see especially pp 132–135. See also Ayer (1956, especially Chapter 3, pp 84–133).

25 The *Abstract* may be found in an edition of the *Treatise* edited by D. Macnabb (Hume 1962 [1739], pp 337–353). There is some doubt about the authorship of the *Abstract* (see pp 31–36). The existence of an *Abstract* was known from Hume's letters, and it was believed that it had been written by Adam Smith at the age of 17. When the *Abstract* came to light in 1933, J.M. Keynes and Piero Sraffa, who edited it, argued strongly that it was written by Hume himself.

26 Mach (1907 [1883]). The book is both an explanation and a critique of mechanics and the two are interspersed throughout the book. See especially pp 216–255 and 481–494.

27 Locke (1975 [1690], Book III, Chapter III).

28 Aaron (1967, p 97).

29 Berkeley (1901, Introduction).

30 Kant (1793).

31 See the quotations and the discussion in Aaron (1967, pp 106–112).

32 Mach (1914, pp 1–45, *passim*, 49 and 310–320). See also the references in note 26. Curiously, Mach traced the idea of conceptual economy to Adam Smith (1914, p 49).

33 Poincare (1914 [1908]).

34 For example, one of the complaints of Hollis and Nell (1975, pp 224–232) against neo-classical micro-economic theory is that 'there are no bearers for neo-Classical variables'

35 Mach (1907 [1883], pp 58–70 and 321–332).

36 Schlick (1974, pp 69, 82–85 and 399). The remark about similarity and psychologists is on p 8 and his critique of Mach on pp 201–209. Carnap remarked that Schlick for a long time regarded himself as a philosophical realist.

37 Wittgenstein (1922).

38 Wittgenstein (1922). In the numbering system which he applied to his statements, the quotations are 1.1, 2.1, 2.171, 2.223. The rest is put together from 4.01, 5, 5.134, 5.1361, 4.211, 4.21 and 4.25. In later life, Wittgenstein more or less repudiated his earlier analysis.

39 Mises (1951, pp 93 and 95).

40 Carnap (1967, pp 158–160).

41 Weinberg (1936, pp 227–288 and especially pp 251–257 and 284). See also Carnap's Preface (1967).

42 The dictum may be found in James (1931 [1907], p 204). James, like Carnap after him, also held that the idea of a pure sensation is an abstraction. See James (1950 [1890], Vol II, p 3).

43 See Ayer (1940, pp 84–92). Ayer takes Carnap to imply that any comparison of a proposition with fact is impossible and that we 'may regard as "true" any system of propositions that we choose' provided that it is 'internally self-consistent'.

44 The falsifiability criterion was first stated in Popper (1959, especially pp 84–111).

45 Popper (1972, pp 7–9, 145–146 and 318; see also pp 67–75 and 241–246).

46 Friedman (1953, pp 3–43).

47 As an example of 'whatever' one could perhaps cite Fair (1978).

48 *The Listener*, Vol. 99, No. 2549, 2 March 1978, p 270. The writer is indebted to Professor L.M. Lachmann, for drawing this interview to his attention.

49 Quine (1953, pp 20–46). The quotations are from pp 42 and 43.

50 Wright (1975 [1957], pp 40–43).

51 Ryle (1949, pp 25–61).

52 See Aaron (1967, pp 75–82 and 94–101).

53 Karl Popper (1972, p 71). 'If it were not absurd to make any estimate, I should say that 999 units out of 1,000 of the knowledge of an organism are inherited or inborn, and that one unit only consists of the modifications of this inborn knowledge.'

54 Polanyi (1958). The quotations, apart from expressions used throughout the book, are from pp vii, 99, 131, 288 and 313. See especially pp 3–17, 69–131, 249–256 and 286–316.

55 Kuhn (1970 [1962], pp 174–210). The quotations in the following paragraph are from pp 181, 2, 5, 7, 9, 190, 3, 5, 6 and 200.

56 Popper (1972, Preface and pp 38–39 and 42–43).

3

General and Particular Judgements

1. Introduction

It seems reasonable to suppose that most economists have had occasion at one time or another to wonder how economic theory relates to what is manifestly happening before their eyes. In this chapter we shall consider such a question in the light of the discussion of the theory of knowledge in the previous chapter. Our purpose will be to appreciate the rationale of the system of thought to be developed in the next chapter. That system of thought will eventually lead us to certain conclusions about the factual or empirical element in economics which were discussed in my *A Realist Philosophy of Economics*. These conclusions, in turn, will enable us to consider more closely how a knowledge of institutionalized conduct fits into our everyday understanding of economic issues and why the presuppositions of micro-economic theory seem to lead to a neglect of such knowledge.

It is necessary first to set the scene. In considering the old question of the relation between theory and fact, we shall not take the point of view of a person who consults a statistical bulletin or who otherwise comes by facts at second hand. Rather, we shall take the point of view of a person who uses his five senses to conclude that such and such is the case. He may simply be a passive onlooker who notices that there a woman is buying a newspaper and over there a man and a salesman are apparently haggling over the price of a motor car. He may be in a factory or a warehouse taking notes before making entries in books of account or before filling in an official statistical questionnaire, but he may equally well be in a supermarket where he notes that the price of butter has gone up again or that a new brand of toothpaste has come on to the market. He may be alternating between sizing up the situation and contemplating his next move while negotiating a business deal, conducting an academic argument or whatever.

In short, we shall be concerned with what may be called first-hand grassroot factual judgements and not with second-hand recorded facts.

Grassroot factual judgements are of course made by everyone every day. However, there are other kinds of judgement. For example, the judgements expressed in the previous two sentences are not grassroot factual judgements, nor is the judgement expressed in the sentence: 'Producers try to minimize their costs and therefore substitute cheaper inputs for more expensive ones whenever they are able to do so.' These judgements differ from the grassroot factual ones in at least one important respect. If one is prepared to express them at all, one can express them in the present tense anywhere and at any time, whether in the midst of a crowd at a fun-fair or alone in a dark room. In this sense they are general whereas grassroot factual judgements may be said to be particular – and in the sequel the word *particular* will be restricted to this sense.

If general and particular judgements appeared to be entirely distinct there would probably be little with which the theory of knowledge could concern itself. That they are not so distinct may be illustrated in the same homely terms as were used above. Let us imagine a man and a young child walking past a construction site, watching the goings-on with interest. We ask the man to tell us what is passing through his mind. He replies that he has gathered from the structure taking shape that this must be the first gasworks to be built in the area for many years and that he has recognized it as one of the consequences of the ever-rising price of fuel oil. We then ask the child what he has seen, and he talks of the big cranes and bulldozers, the black boots and yellow safety helmets of the workmen and so on. That the two have come to such different factual judgements does not surprise us. We do not expect the child to have formed, or to have heard of, general judgements about cost minimization and the substitution of cheaper inputs, about the heat and motion producing properties of gas and oil, about the technology that allows gas to be derived from coal and so on. On the other hand, it does not surprise us much that the child recognized a safety helmet and had thus formed some general judgement, among others, about what happens to people who are struck on the head by a heavy object descending from a great height.

The point of this illustration is that general judgements are in some way involved in grassroot factual judgements, i.e. in our understanding of what is happening in the world around us. That much has often been remarked upon.[1] But there has always been a problem about the nature of the general and of the particular and especially about the relation between them, as the review in the previous chapter has shown. The issue has been considered variously in terms of intelligible form and perceptible flux, quality and substance, the problem of induction, inductive logic of freely created definitions and hypotheses and the facts against which hypotheses are tested. In less philosophical terms the issue has been considered as the relation between intellectual disciplines such as economics and the objects and events of everyday life.

There is no single answer to the question of how this relation is currently conceived with regard to micro-economics. The main founders of micro-economic theory had different aims and different attitudes to the purpose of theorizing – a matter that was reviewed briefly in my *A Realist Philosophy of Economics* – and much in the way of presuppositions was inherited from the older political economy which had itself been exposed to diverse influences for a considerable time. Since various philosophical approaches and ideas have left their traces on the subject, there are also various attitudes to the relation between the general and the particular or between pure theory and experienced events.

2. The natural attitude

As suggested at the outset, an economist may sometimes have occasion to wonder how economic theory relates to what is taking place before his eyes. On such an occasion he does not necessarily formulate his problem as one of a relation between judgements, as was done above, nor as one of a relation between propositions, as is conventional among philosophers nowadays. In other words, he does not necessarily formulate it as a logical problem.

When attention is concentrated on some subject matter, the form in which that subject matter is or comes to be known is not an issue. When one thinks about, say, markets and market relations, one may be making judgements and using concepts but one is not thinking about judgements and concepts. One is thinking about firms and households or individuals trading in markets. That trading, one takes it for granted, is going on out there somewhere, just as out there somewhere bacteria are decomposing organic material and planets are orbiting the sun. In thinking in this manner, one evinces what Husserl used to call the natural attitude. It is the ordinary way of thinking, i.e. thinking in terms of the *real world* composed of the kind of things one has seen or been told about at one time or another. As economists understand them, markets are not, of course, objects on which one can set one's eyes. Reflection would thus seem to lead to the conclusion that a knowledge of specific markets (e.g. the international sugar market or the Zurich gold market) is an inference from a number of particular judgements and this would lead away from the natural attitude to a formulation in terms of judgements or propositions. But economists have not always drawn such a conclusion.

Nevertheless, it is sometimes difficult to maintain the natural attitude. One would not presume, for instance, that one would also find the subject of economics among the myriad things out there somewhere. Since it is not something one would expect to locate in the world (in the sense in which it was thought in the days of Heraclitus and Parmenides that everything had to be locatable) one may wonder what kind of entity the subject of economics is. Having reflected on the matter a little, one may conclude, for instance, that it is a set of theories or models derived by some process of abstraction and capable of representing the essential nature of economic relations or the laws governing economic activity. If then, endowed with the natural attitude and the conclusions of a little reflection, one should come up against the problem of how theory relates to what is manifestly there, one is quite likely to pose it in terms of models and the real world, theory and practice or the necessary simplifications in a theory and the great complexity of the real world.

3. Inductive generalizations and logical truths

We turn now to cases where more than the rather casual thought described above was given to the question of what economists are doing and how it relates to the world around them. Though in many of these cases there was also mention of laws and forces operating in the economy, the problem was usually seen as a logical one, i.e. as one of a relation between propositions or judgements.

A frequent observation in the days of political economy, made both by critics of the subject and by establishment figures such as J.S. Mill and Nassau Senior, was that political economy was a deductive science. By this they meant that it consisted of general propositions derived by logical inference from certain premises which as a rule were also general propositions. Though one does not often hear economics described as a deductive science anymore, the idea is still relevant. Bearing in mind that mathematical equations and inequalities are forms of general proposition and mathematical operations (we are told) a form of logical inference, one can see that a mathematical economic model said to be based on assumptions (i.e. not said to be a hypothesis in the positivistic sense – see below) can be described as a deductive system in terms very similar to those used above. The mode of analysis has changed but its spirit lingers on.

Logical inference is of course a thought process which we can execute without looking at the world around us. If, therefore, we want to find out whether economics as a deductive science has some connection with our everyday experience of the world, we must look for a relation between grassroot factual or particular judgements and those general judgements or propositions which are the premises for deduction or are the assumptions on which models are based. References to the latter in the economic literature are often less than altogether clear. Three examples follow:[2]

> That every person will choose the greater apparent good; that human wants are more or less quickly satiated; that prolonged labour becomes more and more painful, are a few of the simple inductions on which we can proceed to reason deductively with great confidence. (Jevons)

> The observation requisite for the selection of premises may sometimes involve little more than the reflective contemplation of certain of the most familiar of every-day facts. (Keynes)
>
> The propositions of economics 'are deductions from simple assumptions reflecting very elementary facts of general experience'. (Robbins)

On one interpretation of these remarks, the premises or assumptions from which economists reason are inductive generalizations, i.e. general judgements which are conclusions one reaches by adducing many particular judgements after much experience. Economists cannot be criticized for not being more explicit about this procedure since no one has yet managed to give a clear and satisfactory account of induction. But vagueness about inductive procedures makes it hard to tell whether the assumptions of economics really are inductive generalizations or whether they are logical, analytical or a priori truths which, being reached by thought alone, like the propositions of mathematics, do not need and are not derived from an induction (adducing) of grassroot factual or particular judgements.[3]

For instance, the first of the *simple inductions* cited by Jevons (that every person will choose the greater apparent good) looks to be a logical truth, following purely from the meanings of the words. In the ordinary meaning of choice, a person chooses when he selects what appears to him in his present circumstances to be the best of a number of alternatives. A person who appears not to have given any thought to what is best, according to whatever criterion is important to him, would not be said to have made a choice. It may of course be significant to assert simply 'that every person chooses', but again this is not an inductive generalization. An object or a creature which appears never to have made a choice would hardly be called a person. Choosing is a necessary though not a sufficient condition for anything being human. Similar considerations apply to Jevons' third example (prolonged labour becomes more and more painful). Not all prolonged activity becomes painful and, if it does not, one may not be inclined to call it labour.

It is rather strange that Jevons, with his extensive work in the philosophy of science,[4] should have referred in this context to *simple induction*. Perhaps it was merely a manner of speaking. Certainly, some of his other remarks suggest that he did not really consider his examples to be inductive generalizations. In the paragraph preceding the one from which the quotation was taken, Jevons stated briefly but clearly what was to become more than 40 years later the logical positivists' and more or less Popper's view on the problem of induction. But he did not apply this analysis to economics unequivocally. 'The Science of Economics', he said, 'is in some degree peculiar' and one reason for this is 'that its ultimate laws are known to us immediately by intuition'. The word *intuition* does not seem appropriate in the context of inductive or empirical generalizations. Again, he described his theory as 'almost as self-evident as are the elements of Euclid, when the real meaning of the formulae is fully seized'.[5] It is hard to believe that Jevons would have regarded the elements of Euclid, or anything else that is self-evident when its meaning is fully grasped, as 'simple inductions'.

It is plain from the context that the *reflective contemplation* referred to by J.N. Keynes focuses on inductive generalizations. However, when he mentioned the *economic postulates* enumerated by various writers, there were many about which the question raised above about Jevons' simple inductions can also be raised, albeit that the answers are more debatable. The first two postulates given by Cairnes, for example, are 'the general desire for physical well-being, and for wealth as the means of obtaining it' and 'the intellectual power of judging of the efficacy of means to an end, along with the inclination to reach our ends by the easiest and shortest means'. In the same place Keynes also quoted Senior as saying that 'the general facts on which the science of political economy rests are comprised in a few general propositions, the result of observation or consciousness'.[6] Keynes did not comment on the fact that Senior seemed to feel that something other than observation may be involved.

Robbins addressed himself explicitly to the question whether the assumptions of economics are inductive generalizations or 'purely formal relations' and came down on the side of induction.[7]

> It is a characteristic of scientific generalizations that they refer to reality ... they are distinguished from the propositions of pure logic and mathematics by the fact that in some sense their reference is to that which exists, or that which may exist, rather than to purely formal relations.
>
> In this respect, it is clear, the propositions of Economics are on all fours with the propositions of all other sciences.

On the other hand, Robbins indicated that he was presenting the views of the Austrian school of economists and frequently referred to Ludwig von Mises.[8] But von Mises, at least in later years, as we shall see, insisted that economic theory is based on a priori premises, i.e. on logical truths in no way gleaned from (sense) experience. According to him, the propositions of economics are based on the formal relation of means and ends and are indeed similar to the propositions of logic and mathematics. Moreover, it is unlikely that he would have gone along with the implication of Robbins' statement that therefore logicians, mathematicians and economists deal with something that does not exist – implications which are sure to make all rationalists frown. A priori categories, logical truths and analytical statements are of course open to various interpretations and their ontology (i.e. their status in reality) is a contentious issue, as we shall see.

In many cases it is not at all easy to identify the assumptions made in economics as either empirical or analytical or *purely formal*. Often assumptions are qualified by the word *simplifying*. On the one hand this suggests that what is being assumed has not really been found empirically but is something easier for the mind to handle. On the other hand, it suggests that some connection is yet retained for otherwise nothing is being simplified. Sometimes an assumption is supported by an appeal to a type of argument that is also not easily characterized. For instance, the law of diminishing returns is sometimes said to be quite obviously true because: 'If it were not so, then all the corn in the world could be produced from one acre of land.'[9] This appeal may be quite convincing, but it is directed neither to pure logic, because the state of affairs spoken of is not a logical impossibility, nor surely to direct experience, for who has ever tried to supply the whole world with corn from an acre of land. Even Robbins, though he was adamant that the main assumptions of economics 'are based upon experience', admitted that 'there may be room for dispute as to the best mode of describing their exact logical status'.[10]

However, the *logical status* of assumptions is a crucial consideration for the question of how economics as a deductive science relates to everyday experience. If assumptions really are inductive generalizations, the general propositions that make up the subject of economics are ultimately derived from particular propositions, i.e. from reports of grassroot factual judgements. In that case, it is of course still not known how such a derivation is carried out. But even if one accepts that there is no inductive logic, one can still content oneself with presuming, as Hume did (see p 22, this volume), that it is a matter of an association of ideas, i.e. that one notices the constant conjunction of certain elements of grassroot factual judgements. If, on the other hand, the assumptions of economics are purely logical, analytic or a priori, they are associations of ideas that need no empirical prompting and no relation with everyday (sense) experience is established. In that case, it does not necessarily follow that the general propositions that make up economics are in no way related to the grassroot factual judgements that make up everyday experience, but it does follow that it is simply not known whether they are or how they are related.

4. The innovations of positivism

Since the various remarks quoted in the previous section were first expressed, the advent and dissemination of logical positivism has put the whole matter into a new perspective. The anti-metaphysical programme of the early positivists was developed by the logical positivists (and by Popper) into a new attitude towards the problem of induction and this brought with it also a new criterion for distinguishing between synthetic (i.e. empirical) and analytic general propositions (or rather, an old criterion in a new guise). The synthetic general proposition is often rendered in this context by the word *hypothesis*, which may also be understood as a candidate for a general judgement.

Positivistic notions have gained a measure of popular appeal among economists, but the discussion on the realism of assumptions which followed the publication of Professor Friedman's paper on 'The Methodology of Positive Economics' showed that many economists, at the time at least, did not really appreciate the innovations the logical positivists had introduced.[11] Friedman (1953, p 14; original emphasis) probably represented the positivists' views quite well when he contended that 'to suppose that hypotheses have not only *implications* but also *assumptions* … is fundamentally wrong and productive of much mischief'. Unfortunately, he proceeded to create mischief himself by presenting the matter so provocatively that various economists stood up to argue that the non-existent assumptions should be realistic. That such astonishing arguments should have been put forward seems to have been due partly to the fact that Friedman himself reintroduced the term *assumption* and partly to the fact that a hypothesis itself can be understood to consist of one or more assumptions.

In the context of positivism, however, hypotheses do not consist of assumptions or premises in the sense in which the writers quoted in the previous section spoke of assumptions and premises, nor in the sense in which those who insist on the realism of assumptions must think of them, i.e. as inductive generalizations. Positivists, as well as many others, believe that inductive generalizations are impossible, that general propositions simply cannot be composed or logically derived from a number of particular propositions or elementary propositions as they are called in this context.[12] Hypotheses cannot, as it were, be extracted from the physical and social environment, or rather from one's perception of it, and therefore the realism of a specified hypothesis cannot be judged by a quick mental check to see whether it can be justified by such an extraction. A positivist would probably say that it is the vagueness involved in describing assumptions as, for instance, 'reflecting very elementary facts of general experience', as Robbins did, that allows one to gloss over this insuperable difficulty. Furthermore, he would probably object to the expression *general experience*. He may wonder what he is to understand by it when it appears to be a contradiction in terms.

In the manner of Hume's thought positivism holds that a fact of experience can always be described as single and particular (rendered above by *grassroot*) and that one can never experience with the five senses what is asserted by a hypothesis or general proposition such as those of economic theory. The only way to establish a link between a general proposition and reports of facts of experience (i.e. elementary propositions) is to show by deduction that what is asserted by a general proposition logically precludes a certain fact of experience, i.e. hypotheses have factual implications but no factual assumptions. This is the rationale of testing in later logical positivism and in the work of Popper. If a fact precluded by a hypothesis is experienced, the hypothesis is disproved or falsified (with certain reservations). If no such fact is experienced, the hypothesis is of course not proved true in some sense. It is only provisionally valid because not disproved. Interest has centred in precluded facts of experience because, in strict logic, no number of particular cases, however large, could confirm, for instance, that producers always minimize costs, but one particular case could show that they do not always do so.

Friedman changed the emphasis in this analysis somewhat. By using what may have seemed to him the more felicitous word *prediction* for the stodgy *logical implication* and by speaking of *successful prediction* he gave or perhaps wanted to give the impression that the whole analysis is only about mental gadgets for achieving certain ends. For the reasons given above, however, most versions of positivism have concerned themselves only with deducing what cannot happen if a hypothesis is to be regarded as valid and not with predicting what will happen. Quite apart from whatever other difficulties there may be, one could not predict or deduce a particular event (or fact of experience) from a hypothesis or general proposition alone. Some other particular would have to enter the premises. For example, using only the general proposition that producers minimize costs or that water is a compound of two parts hydrogen and one part oxygen, one could not predict or deduce that this or that will happen at a particular time and place.[13]

While few positivists, one imagines, would object to Friedman's practical bent of mind, they nevertheless have often drawn attention to the two rather more academic innovations mentioned at the beginning of this section. They have claimed firstly that they have removed the problem of induction, the enigma of how empirical but general propositions are extracted from fleeting, particular occurrences. The logical positivist says quite simply that they are not extracted. One does not reason from the particular to the general but from the general to the particular. Following on from this, they have claimed secondly that the testing of hypotheses provides a criterion for distinguishing between *scientific* propositions with an empirical reference and propositions which are either analytical (such as those of mathematics) or simply nonsensical.[14] Tests do not actually have to be conducted. A hypothesis falls into the *scientific* domain if it is possible to conceive a test for it, i.e. if one can conceive a fact of experience which would be accessible to anyone who looked for it and which would falsify the hypothesis.[15] Therefore, provided that one can identify nonsense, one may have an armchair way of distinguishing between synthetic (empirical) and analytical propositions.[16] As we saw in the previous section, it is not always easy to make this distinction.

The early positivists tended to make the distinction between scientific and other propositions one between meaning and nonsense, though they exempted mathematics and logic from the latter. Popper spoke of the demarcation criterion for distinguishing between propositions of the empirical sciences and those of other fields but did not make meaning depend on it. Hutchison recommended the criterion as a means of identifying the *propositions of pure theory* in economics, i.e. the purely analytical, non-empirical propositions which he exemplified by: 'Under perfect competition firms are of optimum size.'[17]

5. The facts required for testing hypotheses

Positivists have contributed less to a clarification of the relation between general and particular than they may at first have thought, because they were left eventually with a new enigma, namely: 'Just what is a fact?'. In ordinary discourse *the facts* are often referred to with some reverence and phrases such as 'letting the facts speak for themselves' are used as though nothing could be plainer to the eye than a fact. However, when the logic of positivism is followed up, it soon becomes apparent that facts have to be conceived in such a way that they become extremely elusive.

The difficulties positivism encountered with regard to elementary propositions – the reports of indubitably *given* facts – were discussed at some length in the previous chapter. To remind us of these difficulties, let us compare the facts required for testing hypotheses with the grassroot factual judgements referred to earlier.

The example of a man and child watching a construction site and coming to quite different factual judgements was designed to illustrate that general judgements (for present purposes equivalent to general propositions or tested hypotheses) may be involved somehow in grassroot factual judgements. In this respect the latter differ from the positivists' elementary propositions. A positivist would agree that hypotheses may be involved in ordinary accounts of experience, but he would insist that such accounts must be regarded as interpretations of facts and not as direct reports of facts (i.e. as elementary propositions). He would have to insist on this, among other reasons, because he cannot accept that what is indubitably *given* at a certain time and place can be different for different people, because to his way of thinking the *given* cannot involve hypotheses *brought along* as prior knowledge and because he believes that hypotheses are freely created mental constructs, which is quite the contrary of what he understands by a fact.

In the positivists' scheme, therefore, ordinary accounts of experience or grassroot factual judgements are interpretations of facts. To arrive at the basis of hard fact, or rather at the elementary propositions recording them, one has to purge accounts of experience of all hypotheses used for interpretative purposes. What the residual looks like, few have ventured to say. We have seen that Richard von Mises suggested that 'here pointer at five' and 'there red spot of short duration' are examples. But it is at least open to question whether even these do not also presuppose general propositions.

The elusiveness of the positivists' facts is inherent in the kind of dualism on which positivism is based. Though positivists treat the question of general and particular as a logical problem, there is in more popular versions of positivism (but not, for example, in the logical positivism of Carnap) a juxtaposing of an objective, external, real world and a subjective, internal conceptual realm unreal in the sense of being subject to human whim. (A development of Descartes' *res extensa* and *res cogitans*.) This dualism underlies the notion of facts used for testing hypotheses. As long as this conception is maintained, it will be necessary also to conceive the point of contact between the real world and the conceptual realm at which what is *given* by the real world is *taken* into the conceptual realm, at which facts are, so to say, handed over.[18]

It is not always clear whether positivists' descriptions of theories as filing systems or languages for ordering empirical material (cf Friedman's description, p 27, this volume) imply that the point of contact is located in such a way that the conceptual realm is extended to include all terms expressing the qualities or attributes of things, i.e. the so-called universals. (In technical terms the question is whether positivists are thoroughgoing nominalists – see p 24, this volume.) If it does, the distinction between objective fact and subjective theory cannot be maintained at all, because there is no way in which a fact can be known except in such terms. This last point was already made clear long ago in the distinction drawn by Aristotle between primary and secondary substance or the unknowable *this* and the knowable *this such* and by Kant between the unknowable *Ding an Sich* (thing in itself) and the knowable phenomenon.

The dualist conception, therefore, can be maintained only if it is presumed that beneath our interpretative elaborations, we *take* facts into elementary propositions in the terms in which they are *given* by the real world. In other words, it must be presumed that we abstract essential features which are actually out there in the things of the real world. *Red*, *spot*, *short*, must be presumed to be in all things describable in these terms so that we can abstract them, provided of course that they are not to count as interpretative elaborations. By the same token, it must be presumed that not all of language is in the conceptual realm subject to human whim because the abstraction of essential features implies that at least the meanings of the terms of some basic languages are dictated to us by the nature of things out there in the real world.

This kind of essentialism may seem too fanciful to be taken seriously and positivists, ironically, often look down on it with scorn. But the old empiricist ideas on this are usually no more acceptable to positivists. In Locke's version secondary qualities are produced in the mind by 'the operation of insensible particles' on the sense organs. The primacy accorded by Locke to matter in motion is altogether too metaphysical for most positivists (as it was for Hume), though it may well correspond quite closely to what may be called the view of modern popular science.[19] One can of course avoid both the metaphysics of a stark mechanical universe and that of real essences residing in things, if one quietly abandons, as some of the later logical positivists did, the whole dualism and with it the notion of testing hypotheses against objective facts.

However, the notion is still very much alive in economics and hence also the problem of locating the point of contact between the real world and the conceptual realm, or of drawing the line between abstraction of fact and interpretation of fact. Are *money*, *income*, *capital*, etc. terms of the basic language dictated to us by abstraction and the nature of things or are they concepts that we create in our efforts to interpret or put order into the abstracted terms of that language? It was suggested in the previous chapter that economists who test hypotheses feel free to draw the line between the real and the contrived wherever they find it expedient to do so. In any case, the exigencies of finding statistics usually make it convenient to leave the problem to the people who collect statistics and they, no doubt, do not fret over it either. One cannot of course blame economists for not bothering about red spots of short duration and the like. On the other hand, when one simply gets on with the job without asking too many questions, the distinction between hypotheses and the facts needed to test them becomes entirely arbitrary.

The matter discussed in the last few paragraphs is important enough in the present context to warrant illustration in less abstruse terms. Let us consider an example used by Hutchison to illustrate that the *scientific* procedure, unlike the *philosophical* procedure, allows a decisive settling of disputes through an appeal to facts.[20] (The example is here used for illustrative purposes only. It is not suggested that Hutchison would ever have elaborated it in the way it will be below.)

> Two economists might have an argument as to whether the cheque system did or did not exist in Paraguay. If need be, they could themselves go to Paraguay and investigate, and there is no reason to suppose that they would not soon come to complete agreement as to whether, on their definition of the terms, the 'cheque system' existed in 'Paraguay' or not.

We need not quibble over why this passage is not couched in terms of testing and falsification when that is what Hutchison recommended as a criterion of the *scientific*.[21] Of greater interest is the phrase 'on their definition of the terms'. It was not slipped in casually. Hutchison attached great significance to it.

Let us see first that the role of definitions is here similar to that of general propositions in grassroot factual judgements. Using a mental experiment similar to the one used before, we may suppose one of the economists at some stage to turn to a five-year-old girl at his side and say: 'Is it not plain to you that this person in front of us is effecting payment by sending instructions to his banker ... etc.?' (Let us suppose that this is the definition of using a cheque.) That the girl may look puzzled, as may, perhaps, a herdsman from the mountain regions of Paraguay on being asked the same question, does not damage Hutchison's thesis. *The scientist*, he stressed, 'proceeds by means of the two inextricably interconnected activities of empirical investigation and logical analysis, the one, briefly, being concerned with the behaviour of facts, and the other with the language in which this is to be discussed'.[22] The little girl and the herdsman may simply not have developed the linguistic activity of the economic scientist. Hutchison's reference to inextricability and some of his other remarks on the subject do not remain wholly within the ambit of positivism. We shall return to them. For the moment we may take it that definitions in the positivistic view are in some way involved in more esoteric interpretations of facts.

Let us suppose that the two economists, on first arriving in Paraguay, cannot agree on the facts. As scientists they know that facts are facts and that only their interpretations of facts and their definitions can differ, as is after all not uncommon among economists. They therefore strip their respective interpretations of the facts of all arbitrary definitions and other linguistic conventions and thereby arrive at the basic language dictated to them by abstraction, in which the facts are *taken* as they are *given*. They may then record the facts and perhaps agree that they choose to interpret them differently.

Expressions such as 'effecting payment' or 'sending instructions to a banker' may of course not be reducible to the basic language. In that case, however, they may agree that of the rival hypotheses 'Man scribbles on a piece of paper' and 'Man writes out a cheque' only the latter yields the successful prediction 'Shopkeeper is unperturbed when man picks up object and walks out of shop'.[23] The predicted fact must here be in the basic language so that there can be no mistaking an accomplice in a crime for a shopkeeper or a look of amazement for an unperturbed look. Perhaps the two economists would have to dig yet a little deeper for the facts but eventually they would decide the issue conclusively one way or the other.

Let us suppose instead that the question is whether there is economic development in Paraguay and that a neo-classical and a Marxist economist have gone there to investigate. Like the other two, they also find that they do not agree on the facts. Therefore, they also try to strip their respective interpretations of the facts of all linguistic conventions (that this procedure is unlikely to be acceptable to both parties is irrelevant here) and especially of such troublesome terms as *capital*, *preferences*, *value*, *entrepreneurship* and *exploitation*. However, they still do not agree on the facts. They try to test their rival interpretations but cannot find suitable and mutually agreed on facts for the purpose. Eventually they begin to suspect that the differences between schools of thought in economics are not as superficial as positivists must believe they are and that there is no guarantee whatever in the procedure they are following of ever reaching the common ground of a basic language dictated to them by abstraction, in which it would be impossible to disagree on facts. If they are correct, the notion of testing hypotheses against facts is of no use to them at all in their dispute because the facts they perceive are not independent of their hypotheses.

It has not been the intention here to deny that much agreement on factual judgements is usually reached, nor to point out that it is possible to talk at cross-purposes in economics. The intention rather has been to make the following point: Positivism probably has the strongest claim at present to being regarded as the resident doctrine in economics on the question of the general and the particular. Few economists may know much about its tenets; fewer still may care to defend them. To some, positivism may signify no more than a no-nonsense approach. Nevertheless, many economists embrace positivism de facto, in so far as they see their task as one of devising hypotheses to fit the facts. To adopt this view uncritically, when it is not at all clear whether a definite distinction can be drawn between hypotheses and the facts needed to test them, is surely to build one's foundations on sand.

6. From the natural to the reflective attitude

a. Relation between general and particular

The analyses considered so far have not adequately clarified the problematical relation between the general and the particular or the relation between the theories economists devise when sitting in their rooms and the things they see when looking out of windows. We come now to an epistemological position which, it will be suggested, can provide the desired clarification. It is not the well-articulated and much-discussed doctrine of any one school of thought to which some '-ism' term has been applied, but it may be pieced together from various, especially recent, lines of thought, although its most seminal ideas are not of recent origin. For want of a better term we may, for the time being, call this position the reflective attitude and we may understand by it the position reached when the subjective re-orientation of thought, to which we have paid so much attention, is brought, at least in one sense, to its logical conclusion.

Here we shall restrict ourselves to describing its main features in the broadest outline. Even this will be by no means easy, both because there are no comprehensive compendium-style statements to fall back on and because reflection may seem a deceptively familiar notion and yet involve a manner of thinking rather different from the ordinary. Since a direct statement would run the risk of being quite unintelligible, the explanation will involve a certain amount of talking around the subject. We shall have to approach the subject from various angles (separated more or less into the subdivisions marked by letters) in the hope that by the end of the section the broad features of the reflective attitude will have emerged.

b. Testing propositions

When Hutchison identified science with an appeal to facts, he said that scientists 'have definite, agreed, and relatively conclusive criteria for the testing of propositions' and a little further on he remarked: 'If such intersubjective tests could not satisfactorily be made, there could be no science. A world is quite conceivable in which they could not ... it is simply an empirical fact that over large areas satisfactory agreement can be and is arrived at by such tests.'[24] Had Thomas Kuhn's book been published when Hutchison wrote this, he would surely have realized that he was referring to a 'group-licensed way of seeing' (as Kuhn was to call it) and that science as a social activity presupposes above all a shared paradigm.

But Kuhn's ideas are not within the ambit of positivism. 'What I have been opposing in this book', he said in his Postscript of 1969, 'is therefore the attempt, traditional since Descartes but not before, to analyse perception as an interpretive process'.[25] It is quite possible, according to Kuhn, for different people to perceive different facts at the same time and place. Our example of representatives of different schools of economic thought investigating economic development in Paraguay could have been, in Kuhn's terminology, a case of incommensurable paradigms. But Kuhn was not prepared to accept a complete relativism either[26] (for relativism, see p 35, this volume) and when Polanyi spoke of the *personal participation* of the knower, he did not mean that there is no criterion of objectivity at all, only that it is ultimately not an appeal to facts. However, some remarks on this difficult topic will be deferred to the end of Chapter 5.

Hutchison's remarks show how the puzzling question of the positivists' facts can lead a perceptive mind on to a position which may seem to be a mere variant of positivism but is in fact fundamentally different. Very much the same may be said of Friedman's (1953, p 34; original emphasis) remark: 'A theory is the way we perceive *facts*, and we cannot perceive *facts* without a theory.' The epistemological position implied by these remarks may be expressed as follows: There are no vantage points from which we can survey on one side the hard facts of the real world and on the other the theories (mental constructs, conceptions) by means of which we interpret and order the facts. In other words, if theories are 'the way we perceive *facts*', we are confined to our theories and cannot therefore compare them, taken as a whole, with anything outside and independent of them, as seems ultimately to be required by the notion of judging the usefulness or validity of mental constructs by testing them against facts.

The recognition that we are confined intellectually to our perceptions and conceptions is one way of giving expression to the reflective attitude. It is not a new idea, but it is especially marked in some recent, otherwise quite disparate, philosophical thought (some of which was reviewed briefly in Section 5 of Chapter 2). It is at least a latent implication of Quine's statement that the unit of empirical significance is the whole of science, as it is of the thesis that the roots of science are paradigms. It is a feature of Polanyi's *Personal Knowledge*, of the reflective analysis in the later Wittgenstein's *Philosophical Investigations*[27] and of much of the contemporary analytic and linguistic philosophy which Wittgenstein engendered.

c. The tradition of thought

However, it will be useful for us to see the reflective attitude in the perspective of the long tradition of thought which the previous chapter traced through the centuries. One can discern in that tradition a tendency away from a preoccupation with what there is and towards a reflection on what is known and understood. One could also call it a tendency from cosmology and metaphysics to epistemology or from ontology to logic. It has not of course been an even, step-by-step progression, but rather a tendency that has reasserted itself time and again as though it were inherent, like a logical implication, in the tradition of thought. We shall refer to this tendency as the re-orientation from the natural attitude (as explained in Section 2 above) to the reflective attitude and our task will be to show what it entails and why it is significant.

In the times with which the account in the previous chapter began, it appears to have been taken for granted that whatever one thought and talked about were corporeal and usually animate entities; an attitude which manifested itself in myths and not only in those of the ancient Greeks. We may say that the natural attitude was held in an extreme form. Thus, when the Pythagoreans came to be preoccupied with mathematics, it was not a peculiar quirk on their part that they thought of numbers as the ultimate bodily constituents of all things. When, in Popper's words, Heraclitus 'discovered the idea of change', he groped for a *Logos* which to us now bears a faint resemblance to the concept of equilibrium. But in keeping with his times, he felt obliged to locate it and thought that he had found it in living fire.

We may smile at this nowadays and yet we may be hard put to explain why it is naive. Just such an explanation, however, is needed for our present purpose. We need not deal with the thought of those distant days because the natural attitude is still with us of course. Some cases of this manner of thinking are nowadays characterized, in Whitehead's felicitous expression, as instances of 'the fallacy of misplaced concreteness'.[28] We may consider an extreme case of misplaced concreteness and at the same time give our question a modern tone. Let us imagine a critic of economics who gives voice to the following misgivings:

> You economists say of the co-ordinating price mechanism that some countries have it while others do not, that some used to have it but no longer do. The price mechanism therefore seems to be located in space and time. Yet can you take me to it, can you point to it and say: 'There it is'? Furthermore, you speak of the current state of technology as though it were a describable feature of the world. Yet where is it? Is it perhaps caught in cerebral circuitry or between the covers of books? Again, you speak of consumer tastes or preferences. Where do I find them in heads, in hearts or perhaps, like gut feelings, in bellies?

The fact that we want to say that such questions are silly shows how far thinking has moved over the centuries from the natural attitude, at least in certain contexts. However, it is hardly satisfactory for present purposes simply to say the questions are silly, for it is in arguing the case that the reflective attitude is brought to our attention. But how do we argue the case?

d. Philosophical systems

There is, in fact, no shortage of philosophical systems that could be adapted for arguing that our imaginary critic's attitude is naive and his questions silly. On positivistic principles, for instance, the price mechanism, the state of technology and consumer preferences could be accommodated, like equilibrium and perfect competition, among mental constructs or expedients, in the mental filing system used for ordering facts. One does not then ask of a mental expedient where it is. It is not in the world; it is in thought. However, positivism is only one of many variants in the long tradition of thought following Parmenides. That probably all the variants could be adapted for the same purpose, whatever the intentions of those who originally put them forward may have been, reflects the tendency away from the natural attitude which was spoken of above.

We may consider the Being-Becoming analysis which dominated thought in antiquity. (See Chapter 2, Section 2.) John Dewey, writing on its mature form in the Aristotelian system, put its central tenet as follows: 'Change as such escapes intellectual apprehension. It can be known only in so far as it can be included within fixed boundaries which mark its beginning and its objective end. … Change is known, in other words, only as it is enclosed within fixed limits.'[29] Heraclitus, it should be remembered, had drawn attention to change and to the consequent difficulty of stating what there really is in the world. Parmenides had demurred at the cosmologies of his day, at the accounts of what there really is in the world, and against the changing and variable physical world reached by the five senses he had set what with hindsight may be called the intelligible form reached by reason and reflection. In the intellectual climate of the time, the perceptible, the physical, the changing and variable came to have more or less the same meaning in an appropriate context, namely, the world as taken for granted by the natural attitude.[30] It is a measure of the influence of Parmenides that this world was considered knowable only in so far as, in Dewey's expression, 'it is enclosed within fixed limits', the limits, termini or terms within which it is known being intelligible forms in the manner of the intelligible reality of Parmenides. (The latter's analysis also led to other philosophical positions, such as that of the ancient atomists.) Furthermore, it was not as much a matter of epistemology as Dewey's words may suggest. The fleeting events and changeable things of the real world (as economists are wont to call it now) were caught up, as it were, between terms which were the only real, complete and constant beings or essences. There were variations of this, of course, such as those of Plato and Aristotle, but they had in common that the ontology of the natural attitude had been replaced by the ontology of the intelligible form.[31]

In later thought, characterized by a certain self-consciousness on the part of the thinker (see Chapter 2, Section 3), the perceptible, physical, changing and variable world was separated from the intelligible form, each having its own sphere of existence. In the ontological dualism of the two realities of matter and mind, the physical for the first time came to be opposed to the mental, psychical or ideational, as the intelligible form was now conceived. But the ontological picture was nevertheless varied and complex. In medieval times, it should be remembered, there had been a long debate between nominalists and realists, i.e. between those who saw the intelligible form, in the guise of the universal, as a concept or mere word and those who accorded it a somehow more substantial existence. In subsequent thought, a similar ontological division has persisted.

On the nominalist side the intelligible form has assumed, among others, the guise of the secondary quality and nominal essence of Galileo's and Locke's empiricism, the guise of a model as understood by many economists and the guise of the mental construct or expedient of positivism, of which the representatives in economics are the hypotheses and filing systems of Chicago and elsewhere. In so far as the nominalists' intelligible form is somehow less than real or has an inferior reality (a mere thought), more reliance has to be put on the physical world. But the latter is no longer the world as taken for granted by the natural attitude. Under the influence of the great success of classical mechanics, the physical realm of the old empiricists came to be an austere world of matter in motion and of mathematically formulated laws, while the perceptible, physical world of the positivists, in keeping with the greater self-consciousness of the knower, enters human consideration only as *facts*.

On the realist side the intelligible form has assumed, among others, the guises of the innate ideas of Descartes' rationalism and the synthetic a priori categories of Kant's critical idealism. Among the representatives of these in economics are the a priori categories or concepts (such as *human action* for which von Mises became known and the *real definitions* which have recently been put forward by Hollis and Nell as the staple of what they call the Classical-Marxian tradition in economics.[32] On one interpretation (we shall see there is another), the intelligible form in these guises retains some of its status as ultimate reality or true being that it had in antiquity (hence the name *realist*. But it is not always clear whether the realist also recognizes a physical world and, if so, whether it is the world taken for granted by the natural attitude and, if so also, what its relation is to the intelligible form. Some idealists, especially post-Hegelian idealists, have committed themselves in this respect.[33] They regard the world of the natural attitude, rather as Parmenides did, as something of an illusion. Ultimate reality is Mind, Consciousness or, as in Bradley's idealism, the Absolute with a comprehensiveness reminiscent of Parmenides' original version.[34] It is clear that these idealists reject the world of the natural attitude, but most people find the ontology which they put in its place either very hard to understand or rather quaint.

There is another way of seeing the whole drift and purport of the subjective re-orientation of thought in which it is not seen to adopt either the nominalist or the realist ontology nor the dualism of mind and matter. This interpretation is far more interesting from the point of view of bringing the reflective attitude to our attention by arguing the case against our imaginary critic. Unfortunately, it is also far more difficult to explain and to understand.

Let us approach the matter as follows. In ordinary discourse one may say that a thought is expressed in certain terms. One may also say that something is conceived, understood or even known in such and such terms or in terms of such and such. Though this manner of speaking is quite familiar it hides considerable difficulties. One can notice a certain similarity between it and the Being-Becoming analysis of ancient times as outlined above. Indeed, the current use of the word *term* (in the sense just indicated) has come down to us via logic and mathematics from those times.[35] The ancients took it that the perceptible, physical, changing and variable was known to them only as something enclosed within terms and these they regarded as real, complete and eternal existence. The implications of the current usage, one could perhaps say, is that facts or phenomena are known to us only as something caught up in propositions or in judgements, i.e. only in so far as they are enclosed within terms intelligible to us. That we would substitute 'facts and phenomena' for 'the perceptible, physical, changing and variable' reflects the greater self-consciousness of the knower which has been called the subjective re-orientation. But this substitution creates few problems. It is with the terms (or the intelligible form) that the subjective re-orientation runs into difficulties. Are they merely thought or are they subsistent entities reached by thought, as the ancients took them to be?

This kind of question occupied many philosophers. Many no doubt did take refuge in the dualism of mind and matter and some combination of the nominalist and realist ontologies. But one may read many others of the past and present in a way in which they are not seen to be driving at this kind of solution. Their problem is a difficult one. It is how to regard the intelligible form and how to accommodate it in a thought scheme without making it into a subsistent entity or explaining it as a psychic entity derivative of physical matter according to some psychological principle still to be discovered. Here we may simply say that terms, notions, or whatever one chooses to call them, and the reflection in which they make their appearance are in an important sense non-ontological.

One can find elements of this tenor of thought in the Cartesian and especially in the Kantian and neo-Kantian traditions of thought. They are also evident in Husserl's phenomenology[36] and in the works of Wittgenstein, Polanyi and Kuhn. Some logical positivists were led in the same direction, and, in fact, such a bias seems to be latent in positivism, as we saw in the case of certain remarks by Hutchison and Friedman. In ordinary discourse we would perhaps not say that we perceive a thing in certain terms, but this extension seems to be implicit in Friedman's remark that a theory is the way we see facts and we cannot see facts without a theory. The remark would not have been out of place if made by a neo-Kantian, by a follower of Wittgenstein or by Kuhn. The last-named spoke of a group-licensed way of seeing that may change and in an interesting contemporary discussion of paradigms and the somewhat similar Wittgensteinian concept of a *language game* there appears the sentence: 'Seeing, then, is dependent on training into a language game.'[37]

There is no question here of an unbroken lineage or monolithic school of thought. We are dealing with a tendency of thought which has manifested itself in various forms and in various contexts. However, it is what we want to call the tendency to the reflective attitude.

e. Ontological-type explanation

There can be no doubt that the matter discussed in the last few paragraphs is not easy to follow. We shall have to elaborate it. To this end we shall first describe what we shall call an ontological-type explanation, by which we shall understand an explanation based ultimately on an assertion of what there is. For our purpose, this type of explanation is important only as a contrast to reflection, as something which reflection is not. It may not always be a good tactic to elucidate something by showing what it is not. In this case, however, there are good reasons. First, experience suggests that very many people, if not most, have in mind an ontological-type explanation when they try to explain something, irrespective of whether this involves them in research, i.e. in going about looking for things, or only in reflecting on what they already know. Secondly, what appears to be so extraordinarily difficult to grasp about reflection, as it is to be understood here, is that it is not like an ontological-type explanation.

Let us return to our imaginary critic and his bent for misplaced concreteness. In further conversation he may have come out with the following: 'When you say that consumers have preferences, you imply that there are such things as preferences. So then, explain to me why I cannot go out and touch preferences or look at them.' In taking up this challenge, we may be tempted simply to save the natural attitude by speaking of intangible and unobservable entities which, though we can never lay hold of them, are out there somewhere all the same and are known by their effects. A little thought, however, shows that this is hardly satisfactory. Moliere, in a well-known passage, poked fun at the medical profession of his day for using arguments of this kind. 'Why do some drugs induce sleep?' 'Because they have a soporific quality.' Likewise: 'Why cannot preferences be seen or touched?' 'Because they are intangible and unobservable.' The procedure for this style of explanation seems to have this form: For any question 'Why *x?*' there is always the corresponding answer 'Because *x* happens to be the case.' We may be inclined to call it an empty answer and explanation because it merely asserts what is already recognized in the question. If we do rule it out as a valid type of answer and explanation, we have no assurance that every question has an answer and everything an ontological-type explanation. We shall see later that this is a matter of some significance.

For the moment, however, taking the apparently more usual view, we shall see a different significance in the inadequacy of answering why-questions by asserting that which they inquire into. It may occur to us that we are likely to arrive at empty answers if we are prepared to expand our ontology without limit. With the formula 'there is an *x* with the quality *y*', we may find an appropriate explanation for any event or thing whatsoever if we set no limit on the forms which *x* and *y* may take and if we can avoid incongruity by placing the specific forms of *x* in a whole sequence of separate worlds or spheres. But our belief in what there is may grow more profuse with every question answered and eventually we would surely realize that we are not really explaining anything but are merely cluttering up our ontology. Sensible explanations seem to require some ontological economizing.

We may perhaps take the following view. To inquire into the nature of a thing, i.e. to explain it, is always to reformulate it as something else, as a compound of simpler elements, as a specific case of a class of things, as the consequence of some cause and so on. After such a reformulation, the new entities that have entered the explanation are again reformulated and the process is repeated as often as is necessary on this view, an explanation is brought to a conclusion when it eventually arrives at something which cannot be reformulated as something else because it is what it is, because it is, as it were, ultimate reality. Every ontological-type explanation is therefore based ultimately on certain explicit or implicit assertions, such as: 'There is matter in motion' or 'There is mass-energy, space-time and mathematics' or 'There are people each of whom maximizes the value of a function peculiar to him'. Questions such as 'Why is there matter in motion' or 'Why is there mathematics?' would still have to be answered emptily by 'Because that is how things are', but most explanations would not be pointless. They would reduce statements about a myriad of events and things to a few basic truths. Explanation, on this view, lays bare the reality behind appearance.

f. Ontology of the intelligible form

That was the first step in the elaboration we undertook above. Now we want to see the following. When we speak in economics of deductive models with assumptions (premises) which are or perhaps are not inductive generalizations (Section 3 above) and when we speak of hypotheses (Sections 4 and 5 above), we are introducing the terms in which things are understood or, in other words, the intelligible form (in the guise of models and hypotheses) into ontological-type explanations. Models and hypotheses are then aspects of the reality to which these ontological-type explanations lead or which they lay bare, though there are weaker nominalist and stronger realist versions of this. In other words, implicit assertions that there are models or that there are hypotheses play very much the same role here as assertions that there is matter in motion or that there is space-time. It must be noted – and this is important – that it is not merely a case of asserting the existence of something *in* certain terms (which is the only way anything can be asserted) but a case of asserting the existence *of* the terms themselves.

In the course of arguing with our imaginary critic we may have said at some stage that perfect competition and indifference curves are theoretical constructs. On being questioned further, we admitted that we meant by this that they are parts of models and that we did not object to the (ontological) assertion: 'There are models and there is the real world.' Unfortunately, we have now fallen into a trap set by our critic. We have to admit that a model represents or corresponds to certain things and processes in the real world, not directly like a photograph but in a simplified and stylized way. If, therefore, we assert perfect competition and indifference maps to be parts of models, we must be prepared to also assert the existence of their real-world counterparts, which may well be the price mechanism and consumer preferences. Not only have we got nowhere in countering our critic, but he also starts raising further questions about what a simplified and stylized representation is, i.e. what is a model, and why it should be necessary to introduce even more entities he cannot go out and touch or look at. Thankfully, we can take refuge in positivism. We now say that the price mechanism and consumer preferences, like perfect competition and indifference curves, are mental constructs and that these are like hypotheses. We add that we readily accede to the (ontological) assertion: 'There are hypotheses and there are facts.' A hypothesis, unlike a model, does not represent or correspond to anything but is merely tested against facts. Awkward questions about stylized representation, simplified correspondence or inductive generalization cannot therefore arise. That, we could point out, is what the argument against the need for realistic assumptions was all about.

No doubt our critic could find new objections, but we shall not allow him to transgress the bounds of his usefulness as an expository aid. The point is that the trains of thought we considered in Sections 3 and 4 above presupposed one or other version of the thought-and-the-real-world ontology. If we had considered the intelligible form in some of its other guises, e.g. as essence, a priori category or real definition, the situation would have been much the same. We might have said, for instance, that common experience shows that it is in the nature of a human being to have preferences. Behind this is the implicit and sometimes explicit assertion: 'There are essences so commonly experienced as to be self-evident.' One modern Aristotelian realist among economists has spoken in this regard of *laws of reality* and of axioms (as in *action axiom* 'so broadly based in common human experience that once enunciated they become self-evident'.[38] Again, we might have said that the notion of preference is not distilled from experience but is inherent in the mind which imposes it on experience. Behind this statement, though different interpretations are possible here, may be the assertion: 'There are a priori categories which order a manifold of chaotic sensations into the phenomena of experience.'

In all these cases there appears to be a further hidden presupposition, namely, that we have the benefit of a vantage point from which we can see thought and understanding, alias the intelligible form, as a subsistent entity (in plainer language, as some kind of thing) or, alternatively, from which we can see, in a quite literally detached manner, how we as the objects under observation come by our knowledge. This is fairly clear in the case of real essences or self-evident natures and in the case of a structure of the mind giving shape to a 'booming, buzzing confusion' of sensations (as William James called it, though he changed his mind, see p 39, this volume) of which we do not actually have any experience. But it is also evident in positivism. The falsifiability criterion makes no sense unless it is already accepted that there are hypotheses and facts, where facts are the elusive kind of the positivist and not the ordinarily understood kind. But the proposition that there are hypotheses and facts cannot itself be regarded as a hypothesis requiring testing because its validity would have to be presupposed for the test. It appears to be taken for granted as though it were an indubitable insight into the nature of things, a vision from some privileged position.

The natural attitude has been replaced by the ontology of the intelligible form, as it was in antiquity, and on it has been grafted the dualism of mind and matter with either a nominalist or realist bias. Economists have of course thought about their subject in the milieu of western thought and in their approach to this they have merely followed many historical precedents. But it is not the approach which we want to call the reflective attitude and nor is it the kind of analysis which can clarify the relation between theory and fact.

g. *Reflection versus ontological explanation*

We have tried to articulate a familiar approach to explanation because it is useful to be able to say, in elucidating the reflective attitude, that reflection is not like this familiar type of explanation. We must now consider where the difference lies, though no more than a very rough indication is possible at this stage. A final preamble involving our imaginary critic may help us towards this end.

A number of economists, having ridiculed the critic, decide to set out an answer to him. When they have failed to reach a consensus after much discussion and argumentation, they agree that each should state the basis of his reasoning. It is perhaps not unlikely that something like the following then emerges. Some say there are people each of whom maximizes the value of a function peculiar to him. Others say there are people each of whom chooses freely and acts in an environment of uncertainty. Some speak of mental expedients and facts, some of essential natures and others of concepts, which for them encompass the philosophical terms innate idea, a priori category and real definition. One, standing a little apart from the others, holds forth about forces and modes of production and a string of '-isms' from capitalism and imperialism to commodity fetishism. Finally, a non-economist bystander ventures the opinion that economics in any case deals only with appearances since the only reality behind everything is matter in motion.

Though caricatured here, differences in ontological presupposition are surely not uncommon in economic discussion and, like religious differences, they are not easily resolved. Let us suppose that our group has the will to resolve them. They may sincerely try to understand each other's point of view and may even succeed, only to be puzzled because there can surely be only one basis in reality. They may consider whether they really differ in substance or whether some of the presuppositions may be reduced to others. One of them may argue that *maximizing man* is a mental expedient, a second that both *maximizing man* and *mental expedient* are concepts, while yet a third may point out that *concept* is itself a concept.

If all this may be taken as an allegory of the sort of thinking that led to the subjective re-orientation of thought, our group would eventually begin to realize that they are labouring under a misconception. They have presumed all along that any inquiry into the nature of some matter must necessarily lead to an ontological-type explanation. They have presumed that they have to demonstrate how the thing being investigated is the consequence of, or is determined by, some ultimate, basic factors, or how it is constituted of basic building blocks about the existence of which there can be no doubt. But they begin to realize that what they have actually been doing is to examine their conception of the subject matter, to articulate the terms in which they understand it. In other words, they realize that they have been reflecting and that reflection is not like an ontological-type explanation. In reflecting, i.e. in articulating the terms in which a subject matter is understood, one does not assert or deny the existence of the terms or of the things in the real world which the terms are taken to denote. Rather, since one asserts something in certain terms and since therefore that something is, as it were, enclosed within those terms, reflection lays open to inspection just what is being asserted.

In choosing another illustration of the distinction just stated, we may leave our imaginary critic behind and consider something closer to home, at least for economists. We shall to some extent be anticipating arguments that will be put forward in later, but lacking the intervening analysis we shall only be able to deal with them very loosely. Let us imagine that we are asked the question: 'Why did consumer A buy goods B rather than goods C?' We are told that C was available to A, i.e. on the market and within his budget and credit constraints, but we are not told anything further about A, B and C, apart of course from what is implied in the question. Standard economic theory is not designed specifically to answer questions like this, but we may infer an answer, for instance, from the textbook case of fitting a budget line to an indifference map. On the understanding that mathematical-type determinate solutions in economic models are taken to represent causal-type outcomes in the real world (a topic discussed in Mittermaier, 2023),[39] we may infer the answer that A bought B rather than C because A preferred B to C.

What would we say if at this stage it was pointed out to us that A actually bought C and not B, though the latter was available to him? Since we do not know who A is and what B and C are, it would be unreasonable to say now that A could not possibly have bought C because he preferred B or that A acted irrationally in buying C despite preferring B. The most reasonable response is to say that in the light of the corrected evidence A bought C rather than B because he preferred C to B. Alternatively, we could say that our previous explanation still holds, except that B denotes whatever A did buy and C denotes anything available to him which he did not buy. However, we may begin to suspect that our attempt at an ontological-type explanation has led us to an empty explanation (as discussed in Section 6e). The statement 'A bought B rather than C because he preferred B to C' is true whoever A and whatever B and C may be, but it is no more informative as an ontological-type explanation than the statement that some drugs induce sleep because they have a soporific quality. Any purchase can be explained by asserting the existence of an appropriate preference, but the ontological economizing required by a sensible ontological-type explanation is not thereby practised.

To counter a possible misunderstanding, we would have to point out that the model from which we have inferred our answer does not imply that there is a new preference for every new purchase, but that every new purchase reveals a point in a pre-existing preference field. The ontology on which we rely is not one of ad hoc entities that come and go with every question answered, but one of an underlying stratum of preference fields. The ontology, we would have to admit, is complex, so complex in fact that we cannot hope to know the shapes of all the preference fields but only certain properties common to them all, which we represent in the model as curvature conditions. This is sufficient for the normal use to which the model is put, but in principle at least, we could point out with pride, our ontology enables us, when its other member-entities are also taken into account, to show how each purchase is ineluctably determined by, how it is the inevitable consequence of, what there is.

If at this stage we are asked whether the preference field which led to A's purchase of B still somehow lurks in A afterwards, waiting to be reactivated in the next period, we would have to admit that we are committed to such a view. This then brings us to the point of the illustration. The endeavour to provide ontological-type explanations for the law of demand and, in a wider context, for the process by which prices come about, has committed us to what we shall call a deterministic conception in which the term *preference* is taken to denote a thing-like entity capable of determining an outcome in conjunction with other factors. In contrast, reflection does not commit us to this conception. Reflection is a type of explanation or clarification in which notions such as preferring, determining, being a thing, being capable of and so on, are taken to be terms *in* which we conceive and express a matter so that an articulation of such terms allows us to examine our understanding either of a fact or of that which we do conceive as a thing-like entity.

The deterministic conception alluded to above may of course also be examined in this spirit (i.e. the reflective attitude may be brought to bear on it). A few tentative and rather imprecise steps in that direction may serve our present purpose of bringing out the contrast between reflection and ontological-type explanations (such as theories which explain the determination of prices). Let us consider whether the notion of one thing determining another is really as fundamental to our understanding as an examination of neo-classical price theory may suggest. For this purpose, we shall turn our attention from statements like 'A bought B rather than C' to ones in which at least A and B are further specified; in other words, from what amounts to the general judgement that there are freely choosing buyers to particular or grassroot factual judgements about certain buyers. Instead of 'A bought B' we have, say, 'Joe had a beer at the Red Lion Inn' and the question is how, or even whether, the notion of determining enters this factual judgement. If it is reported to us in the above words, we may not be sure how we are to understand it. In that case we could question Joe. Let us consider three possible replies. He may say that he simply felt like having a beer. If now it is taken for granted that the purchase must be understood in terms of *determining* (with the notion of determining), Joe would be understood to have said that his purchase was the result of (determined by) a momentary whim and this can easily be recognized as an empty explanation. But if this is not taken for granted, Joe may be understood not to have attempted an ontological-type explanation at all but a reflective-type explanation indicating the terms in which he understands his own action. His reply would make his purchase quite intelligible to us even though there is no ontological-type explanation for it and the notion of determining does not enter the matter at all. Similar considerations would apply if Joe replied that he was curious to see the inside of the Red Lion Inn. The main notion now would be that of a means (the purchase of a beer) to an end. To say that his curiosity determined his purchase, or that the end determined the means, would again lead to empty explanations. But if Joe replies, by way of explanation, that it was a hot day, the notion of determining may be said to enter our understanding. A hot environment leads to (determines) perspiration which leads to thirst. However, a thirst does not necessarily lead to the purchase of a beer.

It may be that we know Joe quite well and feel no need to question him in order to understand his having had a beer. The aspect of Joe we may have in mind could perhaps be expressed in terms such as that a disposition to have a beer does indeed lurk in Joe all the time and is reactivated at the slightest provocation. The neo-classical theorist may now point out that the assumptions of price theory must be seen in this context for it seems to make little difference whether one speaks of dispositions, of tastes or of certain contours in indifference maps or preference fields. However, there are important differences between the *tastes* of ordinary discourse and the *preference fields* of economic theory. We shall pay attention to these later, but one difference is pertinent here, namely, that we do not ordinarily understand a disposition or a taste for some consumer good to determine an action.

To see this, we may consider a different purchase, say, 'Joe bought a bicycle'. We may know Joe well enough not to understand him to have bought a bicycle for the sheer joy of it, as may be the case when he has a beer. We therefore understand the bicycle to be a means to something else close to his heart. As economists we would perhaps understand Joe to budget and plan his future carefully and to reject all those plans (combinations, bundles or baskets of goods and services) which do not include a certain rather costly education for his children as well as all those which would involve his falling into debt. On account of his meagre means, we would understand these preferences to leave him only with plans which, amongst other things, require him to travel to work by bicycle rather than by car. Non-economists would perhaps express all this by saying that the education of his children means a great deal to Joe, that he adheres to a code of conduct not to fall into debt and that he has to cut his coat according to his cloth. It may be that we know that, though the problem of scarcity does not leave Joe untouched, it does not touch him to the extent that he cannot buy a bicycle without bearing in mind all his other planned purchases. We may instead understand him to have had in mind his state of health and the beneficial effects he believes exercise would have on it or perhaps we would understand him to have been influenced less by a concern for his health than by the fact that many of his friends and others have taken to this form of exercise. Obviously, there are many possibilities. The point is that other people's concern for their health, for the education of their children and so on, their fads, fashions and codes of conduct (some of which, when applicable in very many cases, may be called institutions) play a similar role in understanding here as a taste for beer did in the previous example.

We shall investigate the similarity later. What is important here is that neither tastes and dispositions nor institutions normally enter grassroot factual judgements as determinants. If they did, we would have to believe that, had Joe become self-conscious in the course of his purchase and been able to see himself rather as we could see him, he would have watched helplessly how ineluctable forces made him buy a bicycle (determined his purchase). Perhaps some people do understand others in this way, but it seems doubtful whether the majority of neo-classical price theorists do actually understand in this way the actions of their family, friends and colleagues. The notion of determining or at least of contributing cause and effect does of course enter most factual judgements in some way. In our example it enters our understanding of how bicycles and motorcars respectively are propelled, of the connection between physical exertion and the healthy functioning of the body and so on. But it is only one notion (or perhaps complex of notions) among many. On reflection, it is at least questionable whether it is so fundamental to our understanding as to warrant the reification of other notions such as preference. In other words, it is questionable whether it is any more plausible to say that preferences are entities capable of determining something than it would be to say that determinations are entities capable of preferring something.

h. The reflective attitude and transcendental subjectivism

We have now paid some, perforce very loose, attention to what is meant by reflection. We still have to say something about the reflective attitude, for clearly these terms are not synonymous. As the examples just considered were meant to show, we can reflect on the most mundane factual judgements as well as on the presuppositions of a rather esoteric theory. Likewise, we can reflect on our understanding of anything else. It is, however, a rather special case when we reflect on the terms in which we reflect, when we articulate the terms in which we understand our own understanding and examine our conception of conceiving. Out of such reflexive reflection comes an understanding of the extent and limits of our knowledge and understanding. When we approach a subject matter with such an understanding of our intellectual limitations, we approach it with the reflective attitude and, in the terminology used here, the subjective re-orientation of thought is the tendency from the natural to the reflective attitude.

At the beginning of this section the understanding of our intellectual limitations was alluded to as the recognition of being confined to our perceptions and conceptions. The confines are of course metaphorical. We could equally well have spoken of the inescapable truism that the only real world we know is the real world we have conceived. (From which it would follow that an inquiry into any feature of the real world, such as the price mechanism, must be understood as an inquiry into our conception of it, however much observation and experimentation the enquiry may entail.) It may be thought that a mere truism, stating little more than that we know only what we know, can hardly be of much significance. But it is one of many possible, all more or less imperfect, ways of giving expression to the gist of the subjective re-orientation of thought.

Kant approached the matter by pointing out that the words 'I think …' may be prefixed to anything anybody ever says and he then proceeded to draw important conclusions from this.[40] Kant also remarked upon the ability of human beings not only to think but also to say to themselves that they think.[41] In the epistemology of Kant and in the phenomenology of Husserl the analysis which proceeds from this self-consciousness of the thinking and knowing subject is called transcendental.[42] While transcendental philosophy is relevant to our purpose, its quite formidable difficulties make it impracticable to deal with it here. We are of course interested in the matter only as a means to an end in economics, i.e. only in so far as it has a practical bearing on our understanding of the economy and economic events. (That it has such a bearing is also something we want to show.) We shall of course be able to draw on the ideas of various thinkers and especially on certain remarks made by Wittgenstein and Polanyi which seem to come to grips with the issues of transcendental subjectivism without a formal transcendental analysis.

i. Clarifying potential misunderstandings of reflection

We have made rather extensive use of argumentation by example and illustration and this, it may be felt, is a poor substitute for plain statement and rigorous proof. The latter, however, is possible only when one operates within a well-known and widely accepted system of thought or paradigm, if we may borrow the word. It is not possible when a paradigm itself is to be described. In the absence of some universal paradigm in whose terms all others may be rigorously defined, one cannot state a system of thought or paradigm but, rather as Wittgenstein taught in the *Tractatus*, one can only show it. Nevertheless, illustrations do have their limitations. For this reason, it may be best to mention a few likely misapprehensions of what we have called reflection and the reflective attitude.

First, it may be thought that the statement that the only real world we know is the real world we have conceived implies that the world is 'all in the mind', that each knower creates his own world and that the self is the only thing truly known to exist, or again that it implies the view, attributed to some idealists, that in the final analysis there is only mind or ideality. If the statement did have such ontological presuppositions, the change from the natural to the reflective attitude would be merely a change from one ontology to another: the existence of different things would be presupposed. But we went to great lengths to show that this is not what the change amounts to. In reflection one does not make assertions or denials nor does one imply them but rather one analyses them.

Unfortunately, that is not the end of the matter because there appears now to be a contradiction. Mention was made, in connection with the reflective attitude, of the extent and limits of knowledge and understanding. In fact, this is what was meant to be expressed by the statements that the only real world we know is the real world we have conceived and that we are confined to our perceptions and conceptions. Statements about intellectual limitations appear to be about the nature of things, i.e. they appear to be ontological. Yet they were said to be a conclusion drawn from reflection on the terms in which we reflect and this, as we have just reiterated, is in an important sense non-ontological. To show that there is no contradiction here, we would have to go into the formidable difficulties of transcendental analysis, which would be a very big task. However, to counter the impression that a bit of esoteric terminology is being used as a convenient means of disposing of an awkward point, a very rough indication of the difficulties involved will be given.

When we say to ourselves that we are able to know something (i.e. reflect on our knowing), we invoke the correlative terms of knower and known, of cognizing subject and cognized object, such as the self of the observing economist and the object of his attention. Thus, we understand presuppositions about the nature of things in the correlative terms of the subjects who know and the known factual objects. But a difficulty arises when subjects speak about their own intellectual limitations because the subjects who know are then also the known factual objects. One can of course know something about knowers and even about oneself as a knower, but when the knower becomes the known there must always be a further implicit knower, the transcendental ego as it has often been called, because the terms *knower* and *known* always imply each other and make no sense on their own. When one speaks of intellectual limitations one may not have in mind the factual limitations of factual objects who know something but the limitations of the knowing subject or transcendental ego. These limitations do not refer to a factual circumstance but rather to the purely formal circumstance that the knowing subject cannot become the known object without ceasing to be the knower in question. How inferences from such meditations can have a practical bearing on our understanding of the economy and economic events is a matter that will be argued later.

Some appreciation of transcendental subjectivism is needed also for dealing with another likely misapprehension, namely, one that would misconstrue reflection as a kind of psychological analysis of concept formation. A parallel may here be drawn with the debate on so-called psychologism, i.e. the attempt to explain, for instance, the foundations of arithmetic by investigating the formation of concepts. For 19th-century empiricists psychologism was a central tenet (though they did not use the term). J.S. Mill, for instance, held that even the proposition $1 + 2 = 3$ is a fact learnt from such experiences as seeing three pebbles in various configurations.[43] Others, notably Frege, took exception to psychologistic explanations of concepts in arithmetic and logic not only because the explanations seemed to be inadequate but also because the whole approach seemed to be mistaken in principle. Husserl also, after his early work on the philosophy of arithmetic, purged his work of the taint (as he saw it) of psychologism. But he and Frege were led in the direction of a position usually called Platonism and this they did not want to assert either.[44]

The question is therefore also not an easy one. We may have recourse again to the correlative terms of a subject knowing an object in order to give another rough indication of what is involved. The view of psychology underlying the criticism of psychologism is that of an empirical science which tries to give ontological-type explanations, inter alia, of the formation of concepts. In other words, concepts and their formation are here the objects known to the subject. In reflection, however, terms (or if one likes, concepts or notions) are not the objects known to the subject but rather, as it were, the means by which the subject knows the object. If therefore psychology is understood as a science concerned with thought processes considered as events in the world (it is not suggested that psychologists necessarily take this view of their subject), then reflection has nothing to do with psychology.

Another misapprehension may arise from the usage of the word *conception*. One may say, for example, that there are various conceptions (or models) of the price mechanism and that one's own conception of it has changed (or that one has changed it). Since one does not mean that the price mechanism has changed but only one's conception of it, this may suggest that there are conceptions as well as that of which they are conceptions. The metaphor of being confined to conceptions may then be taken to imply that we are somehow prevented from knowing anything as it really is in itself. However, the alternative idea that we can know our own conception or model of the price mechanism as well as the price mechanism as it really is (such as when we say that perfect competition is an unrealistic conception) is also rather strange. It implies that we can know something quite apart from having a conception of it, so that we could also know something of which we have no conception whatsoever. The words *knowing* and *conceiving* are surely used in such a way that this is logically impossible.

The difficulty illustrates the importance of stressing that reflection is in a sense non-ontological. If terms, notions, conceptions, models, sense impressions, etc. (i.e. the terms in which we reflect) are regarded psychologistically as things or events in the world, we must eventually come to the conclusion that they are also the only things or events in the world, or at least the only ones to which we have access. The awkward conclusion is due to us regarding the terms in which we reflect as objects known to the subject instead of as the subject's knowing of the object. More loosely, one could also say that we do not know both our own current conception of the price mechanism and the price mechanism itself, but that our own current conception of the price mechanism is our knowledge of the price mechanism.

Finally, it may be thought that an important trend in western philosophic thought has here been represented as a mere antipathy to stating what actually is the case and as a predilection for saying how things are understood. Moreover, it may be thought that unless statements about how things are understood presuppose ontological commitments which are in fact not made here (such as that reality is Mind, that we create our own world or that we do not have access to things as they really are) then they are substantially equivalent to statements about what actually is the case, the difference between them being merely verbal. The misapprehension here is that statements about how things are understood or conceived are to replace statements about what is actually the case. Such a substitution would in fact be impossible. As a logician might put it, one is concerned with different orders of statement. If statements about how things are understood, or conceived are called p^2-statements, then p^2-statements are always about other statements – let us call them p^1-statements – so that p^2-statements cannot replace p^1-statements. Thus, 'It is a fact that people have preferences' and 'That people have preferences is an assumption' are p^2-statements about the p^1-statement 'People have preferences'. They describe how the p^1-statement is understood, the first of the two p^2-statements stating that the p^1-statement is understood as a statement about what is actually the case.

In Chapter 5 we shall have a closer look at the logicians' device being used or rather adapted here – the device of so-called metalanguages. It is capable of considerably clarifying some issues, but we shall see that it is not without difficulties of its own. It is also capable of some tongue-twisters. For example, 'Statements about what is actually the case are p^1-statements' is itself a p^2-statement and similarly 'Joe bought a bicycle is a p^1-statement' is a p^2-statement. In terms of the device used in the preceding paragraphs; p^1-statements are about objects known to or conceived by subjects whereas p^2-statements are about *how* objects are known to or conceived by subjects. Hence, p^2-statements always presuppose p^1-statements so that they cannot replace them. In other words, we always reflect on something. It would be quite mistaken, therefore, to suppose that a substitution of statements has been proposed. Indeed, the whole analysis is being put forward as the necessary groundwork for showing that neo-classical micro-economic theory is formulated in such a way that economists forgo the opportunity of coming to many statements about what actually is the case, including those which in a wider context would be said to be about institutions.

j. Analysing statements about economic theory and reality

The above analysis of statements about statements may be applied briefly to the question posed at the beginning of this chapter, namely, the question of how the propositions of economic theory relate to what is manifestly happening before our eyes. Let us consider two statements, say, 'People have preferences' and 'Joe had a beer' and call them p^{1g} and p^{1p} respectively. In explaining what we mean by (or how we understand) these statements, we may, by way of emphasis, add to p^{1g} the words 'They really do' and to p^{1p} the words 'He really did'. We could also express this more formally in the p^2-statements 'p^{1g} records a factual judgement' and 'p^{1p} records a factual judgement'. Let us call these p^2-statements $p^{2.1g}$ and $p^{2.1p}$ respectively.

However, the words *factual judgement* do not seem to be used in the same sense in these two p^2-statements. Some reflection early in this chapter already led to the vaguely formulated conclusion that we can distinguish between general and particular (or grassroot) factual judgements. Along these lines, p^{1g} could be said to record a general judgement and p^{1p} to record a particular judgement. The difference begins to manifest itself when $p^{2.1g}$ and $p^{2.1p}$ are expressed informally. For instance, it may be possible, depending on circumstances, to express $p^{2.1p}$ as 'Joe really did have a beer; I was there and saw it with my own eyes'. But while $p^{2.1g}$ may be expressed as 'People really do have preferences', it would never be appropriate also to add 'I was there and saw it with my own eyes'. It appears, therefore, that there is scope in this case (as in many others) for stating how we understand the p^2-statements which state how we understand certain p^1-statements. In other words, to investigate the difference between general and particular factual judgements more rigorously, we have to make p^3-statements about p^2-statements such as $p^{2.1g}$ and $p^{2.1p}$.

Only when that has been done can an answer be given to the question of how the propositions of economic theory relate to what is happening before our eyes, for it will be seen that this is a question of the relation between general and particular judgements. Likewise, our imaginary critic could be answered, because a little thought would show that he asked very much the same question. We shall not return to this critic's misgivings – he was, after all, only an expository aid. But we were concerned in *A Realist Philosophy of Economics* with the way economists have variously understood the general judgement that people have preferences. There I tried to show that the dominant conception at present is such that it leaves little room in economic theory for the many other general factual judgements made in everyday life, including those which, in the manner of thinking here called the natural attitude, are said to be about institutions.

Notes

1 Opposite the first page of the text of *Studies in Philosophy, Politics and Economics* (Hayek 1967), Hayek placed a quotation from Goethe to the effect that everything factual is already theory, namely, '*Das Höchste wäre zu erkennen, dass alles Faktische schon Theorie ist.*' The idea is perhaps the major thesis of the Kantian tradition in philosophy. Popper also insists on it and Friedman mentions it as an aside, even though it seems to create puzzles about testing procedures.

2 Jevons (1879a, p 18); Keynes (1917, p 229); Robbins (1935 [1932], p 104).

3 It was pointed out in the previous chapter that the distinction between factual and logical truths, or rather between synthetic and analytical statements, has been questioned in more recent years. The traditional distinction is taken for granted in the positivistic view on the relation between general and particular in economics (in e.g. Friedman's contention that theory may be viewed 'as a filing system for organizing empirical material'). Since

the intention here is to show how positivists tried to improve upon earlier views on this relation, the traditional will also be taken for granted here. However, the question will be taken up again.

4 Jevons (1874).

5 Jevons (1957 [1871], p 21).

6 Keynes (1917, p 244).

7 Robbins (1935 [1932], p 104).

8 Robbins (1935 [1932], for instance, p xv f). A footnote to the first statement of his celebrated definition of economics (p 16) lists the works of various authors of the Austrian school, including von Mises.

9 Robbins (1935 [1932], p 77).

10 Robbins (1935 [1932], p 80f).

11 Friedman (1953). For examples of the discussion, see the comments by Nagel and Samuelson (1963). See also Hollis and Nell (1975, pp 196–201) and the discussion on falsifiability in economics and the bibliography on the controversy over the realism of assumptions in Blaug (1968, pp 666–675 and 682f).

12 One basis for this belief is that general propositions usually assert that something is true of all, every or any member(s) of a class, which is not asserted by any one elementary proposition and therefore cannot be derived from a number of them either. If this were expressed more rigorously, it would be what is called the problem of induction.

13 Since the text was written, there has appeared Lawrence Boland's comprehensive paper 'A Critique of Friedman's Critics' (Boland 1979). Apparently because Friedman saw the testing of hypotheses as successful or unsuccessful prediction, Bolan treats Friedman's *positive economics* as a case of *instrumentalism*, a term usually reserved for the philosophy of John Dewey. On first reading, Boland appears to accord Friedman's paper a kind reading, but after a closer examination one has a different impression. Since Friedman claimed to describe a positive economics and not an instrumentalist economics, neither *Dewey-type* nor *sui generis*, one has to presume that he simply did not understand positivism and that, in describing what he thought it was all about, he arrived at a different position. Furthermore, since Friedman's *instrumentalism*, as interpreted by Boland, seems to say little more than that one should be very glad if one is lucky enough to find a good rule of thumb, there appears to be very little that one can say for it. In fact, that seems to be the conclusion Boland reaches, albeit in a veiled way.

14 In this context it is always assumed that one is dealing with what economists call positive as opposed to normative statements. Though a positivist would hold that statements such as 'Unemployment is a greater social problem than inflation' or 'Pride and envy are sins' are not testable (unless rather arbitrary definitions are assumed), he would not regard them as analytical either and only fanatical outgrowths of positivism would regard them as meaningless or nonsensical. In a positivistic way of thinking, they simply do not assert or deny any predicate but express exhortations, wishes, commands or the like.

15 See, for example, Hutchison (1938, p 9f). The criterion is really an old one, namely, that a contradiction of a logical truth is unthinkable.

16 It has sometimes been pointed out that it does not seem to be possible to conceive of a test for what is probably the most common example of a proposition in books on logic, namely, 'All men are mortal'. Should one therefore regard it as analytical (or definitional) or perhaps as nonsense? One is reminded here of Shackle's remark that time is alien to reason.

17 Hutchison (1938, pp 23 and 26–27; see also p 6). In the 1960 Preface, Hutchison said that he still regarded this criterion as 'useful or clarificatory' (p viii). Friedman, as we saw,

also recognized the distinction. 'Viewed as a language, theory ... is a set of tautologies.' 'Viewed as a body of substantive hypotheses, theory is to be judged by its predictive power.'

18 Ross (1970) identifies at least three types of theory about the *given* in experience. He argues that they are based on different metaphysical ideas about the nature of thought so that the proponents of these theories are often at cross-purposes. He remarks (1970, p 155) that their dispute is given a 'semblance of a genuine quarrel' by the fact that 'whatever else the theorists are trying to point out when they say that sense-data, objects, or immediate experience are given, they also mean that these are the sole representatives, at the perceptual level, of Reality – ultimate metaphysical Reality. Thus, when they say that something is given, they mean that this constitutes our contact with reality'.

19 There is, however, a philosophical position known as *physicalism*, which is held by some positivists. According to physicalism all significant statements of fact can be reduced to (or translated into) the language of modern physical science.

20 Hutchison (1938, p 7).

21 Since the two economists are arguing about a '*cheque system*', which presumably they do not expect to perceive as a single 'fact', one might expect them to consider the proposition 'The cheque system exists in Paraguay' as a hypothesis to be tested. But there is a technical difficulty about falsification in this case. The proposition could be rendered as follows: 'All or some people in Paraguay sometimes use cheques.' The difficulty attaches to what logicians call the quantifier 'some'. Its inclusion makes it impossible to conceive of any single experience which would falsify the proposition.

22 Hutchison (1938, p 9).

23 Hutchison did not, of course, speak of successful prediction. The positivism of his 1938 book is quite orthodox and rigorous. One exception to this, in which he could be said to have gone beyond positivism, will be noted later.

24 Hutchison (1938, pp 7 and 9).

25 Kuhn (1970 [1962], p 195).

26 Kuhn (1970 [1962], pp 205–207).

27 Wittgenstein (1953).

28 Whitehead (1925, p 75ff).

29 Dewey (1938, p 84f).

30 The word physical comes from the Greek word for *nature* which in turn was derived from a word meaning to *produce* and ultimately from a root meaning to *grow*.

31 The word ontology is used here and in the sequel in the sense of a set of presuppositions or beliefs about what ultimately is real. This usage is found, for example, in Quine (1953) and has become more common in recent years. In this sense one may speak of many ontologies. In the more traditional usage, ontology means the study of existence in the abstract or of being as such and, in this sense, there is only one ontology.

32 Hollis Nell (1975, pp 177–181).

33 See Ewing (1961). Though he considers idealism to be difficult to define, Ewing says that most idealists are united in rejecting the ultimate reality of physical objects.

34 Bradley (1893).

35 See the note at the end of the article on the noun *Term* in the *Oxford English Dictionary*.

36 Husserl's thought evolved constantly and apparently rather self-consciously. His contemplation of the problem alluded to in the previous paragraph is well-recorded. From the discussion of Husserl's manuscripts in Broekman (1963, pp 6–27), it appears that he formulated his first solutions to the problem between about 1907 and 1913. A good survey (from the point of view under discussion here) of the development of Husserl's thought may be found in Sokolowski (1964, especially Chapter IV). Mohanty

(1964) begins with the sentence: 'There is one dominating interest which runs through all the works of Husserl, from the earliest to the latest, and imparts to his philosophical career an almost tragic note: this is the search for a stable via media between Platonism and Anti-Platonism.'

37 Phillips (1977, p 101).

38 Murray N. Rothbard in Dolan (1976, p 24f).

39 Without this understanding, it would be hard to see why economists concern themselves with such questions as false trading, information flows, expectations ('rational' or otherwise) and uncertainty.

40 Kant (1793, Transcendental Logic, para 12).

41 Quoted in Broekman (1963, p 147). The actual quotation is: 'Dass der Mensch nicht allein denkt, sondern auch zu sich selbst sagen kann ich denke macht ihn zu einer Person.'

42 Broekman (1963) is a comprehensive explanation of Husserl's idea of transcendental analysis and its relation to Kant's. The text is in German. Important articles on transcendental philosophy in English, many of them translations, may be found in Elliston and McCormick (1977). See, especially, L. Landgrebe, 'Phenomenology as Transcendental Theory of History', I. Kern, 'The Three Ways to the Transcendental Phenomenological Reduction in the Philosophy of Edmund Husserl' and J.J. Kockelmans, 'Husserl and Kant on the Pure Ego'.

43 Mill (1843, Book II Chapter 6 paragraph 2 and Book III Chapter 24).

44 See, for instance, Elliston and McCormick (1977), the articles by G. Frege (1894) 'Review of Dr. E. Husserl's Philosophy of Arithmetic' (first published 1894) and D. Willard, 'The Paradox of Logical Psychologism: Husserl's Way Out'; Sokolowski (1964, pp 18–24); Mohanty (1964, Chapter III).

4

Subjectivism in Economics

1. Introduction

The question of how economic theory relates to fact – to what is manifestly happening before our eyes – is the age-old problem of the relation between the general and the particular. We have seen that the groping attempts over the centuries to find a solution to this problem have evinced an ever-greater self-awareness on the part of the analyst. We have called this the subjective re-orientation of thought. The question now arises whether this subjective re-orientation is in any way related to the introduction of a subjective value theory and a theory of demand into economics a century ago – a development sometimes referred to as the subjective revolution in economics, though more commonly as the marginal revolution.

Superficially at least, there are some quite obvious parallels. 'In the subjective revolution of the 1870's, the first step in the direction of subjectivism was taken when it was realized that value, so far from being inherent in goods, constitutes a relationship between an appraising mind and the object of its appraisal.'[1] Again, Berkeley's New Principle (*esse* is *percipi*- to be is to be perceived) found a new and limited application. To be valuable is to be perceived as valuable. Since virtually everything economists study involves value or prices in some way, it could be said that economic reality, like the reality of some idealists, does not exist independently of minds.

However, we shall find that, with the exception of what may be called the Menger-Mises line of the Austrian school, the *new* economics stemming from the 1870s had very little to do with the subjective re-orientation of thought. The picture which will emerge may be summed up as follows.

Demand theory introduced the notions of usefulness and pleasure or satisfaction (more or less fused into the concept of utility) into economics, though they had not been neglected altogether by the older political economy.[2] But utility was introduced into economics in a scheme which was inspired by the example of classical mechanics. Utility was regarded as one of the determinants of price, in the sense in which the independent variables in equations are the determinants of the dependent variables – which is also the sense in which determinants and determination may be said to be a feature of classical mechanics. The notions involved in this aspect of mechanics proved to be incompatible with the notion of utility as it had come to be used in social theory over the centuries. As it turned out, utility and its associated notions, such as pleasure and value, were incorporated in marginalist theory in a way which in effect eliminated them or at least rendered them quite otiose.

Menger's thought, however, ran along quite different lines – it was distinctly Aristotelian. He realized that the concepts of mathematics and physics could crowd out the more purely economic concepts and eschewed the path followed by other marginalists. In Menger's view, the task of the economic theorist is to find exact laws. The nature of such exact laws is not entirely clear – von Mises later tried to clarify and update them by rendering them as neo-Kantian a priori. We shall try to understand them by looking at Menger's attitude towards free will, error and ignorance, which in his day constituted a problem area in economics very similar to that which nowadays goes under the heading of uncertainty.

Merger's exact laws and exact science follow from what he called a certain direction of cognitive endeavour (*Richtung des Erkenntnisstrebens*). He thus evinced the self-awareness on the part of the analyst, which we have called the subjective re-orientation of thought. He, like many before him, did so in an effort to deal with the problem of the general and the particular, though neither he nor von Mises had a really workable solution to the problem.

This subjective aspect of Menger's thought has, however, nothing to do with the subjectivism which is a feature of the modern Austrian school in economics. The difference may be indicated in the following way: Oskar Morgenstern once remarked that it is misleading to speak of a subjective theory of value. One should speak instead of a theory of subjective value.[3] In other words, the subjectivism of value theory is the subjectivism of the valuing individual or, as we shall call it, the subjectivism of subjective objects (as opposed to the subjectivism of the self-awareness of the analyst). This subjectivism may be preferable to the quite otiose rendition of subjectivism in neo-classical theory. Nevertheless, when its implications are followed up, it will be seen to lead to a dead end – one, moreover, against which Menger expressly warned his readers. The subjectivism of subjective objects holds that the objects of economic study have the property of being subjective and this quickly comes to mean that they are capricious, intractable and beyond the reach of a determinants-based theory or, for that matter, of any rigorous theory. Much of the work of the Modern Austrians has therefore tended to be a negative iconoclasm, a nagging of neo-classical economists on the grounds that neo-classical theories are far removed from what in an everyday sense are regarded as economic questions.

In contrast, the Menger-Mises line of thought – from which the strictures on neo-classical theory also follow, in fact more rigorously – contains the germ of an idea from which could grow an approach to economic questions which would be a positive alternative to the neo-classical economist's preoccupation with finding determinants – a preoccupation which, it may be argued, has hamstrung economic theory in this century.

2. Utility and mechanics

In the 'new' economics which took hold after the 1870's, utility became one of the determinants of price. Utility could serve this function because many of the early marginalist writers brought it into economics in a scheme inspired by the example of mechanics. Three of these writers are considered below.

a. Jevons

Jevons first announced his proposals for a new kind of economic theory in a paper read to the British Association in 1862.[4] Paragraphs 2 to 9 of that paper were quite clearly inspired by Jeremy Bentham's teachings; even the turn of phrase was distinctly Benthamite. In the full-scale exposition nine years later and in the second edition of that work 17 years later, the utilitarian influence was still very much in evidence. 'The theory which follows is entirely based on a calculus of pleasure and pain' and it 'may be described as the mechanics of utility and self-interest'. Jevons described the 'words of Bentham' as *grand* and *full of truth* and devoted the second chapter to the theory of pleasure and pain. On reading the Preface of 1879, one gets the impression that Jevons' enthusiasm for a mathematical formulation, which had of course been a factor from the start, had by that time pushed his interest in the utilitarian basis into the background. But he still acknowledged: 'As to Bentham's ideas, they are adopted as the starting point of the theory given in this work.'[5] In an article published in the same year, Jevons also criticized J.S. Mill for trying 'to give geniality to Utilitarianism' and advised his readers: 'We must hold to the dry old Jeremy.'[6]

How Bentham's utilitarianism should be interpreted is a question open to debate, as Section 3 below will show. But there can be no doubt that Jevons really did interpret it as a *mechanics of utility*. His book can leave no other impression than that his theory was conceived, as Jaffé has put it, 'on the analogy of a mechanical balance of physical forces'.[7] In the original preface Jevons said: 'The nature of Wealth and Value is explained by the consideration of indefinitely small amounts of pleasure and pain, just as the Theory of Statics is made to rest upon the equality of indefinitely small amounts of energy.' He spoke of pleasures neutralizing pains and, at one point in the text, he was struck by a similarity even in a matter of detail: 'It is curious, moreover, that, when we take the theory of the lever treated according to the principle of virtual velocities, we get equations exactly similar in form to those of the theory of value as established above.'[8]

b. Walras

According to Jaffé, Walras disliked utilitarianism.[9] His inspiration came instead from his father, Auguste Walras, from his father's one-time classmate, Cournot, and through them from J.B. Say, Condillac and Turgot as well as Galiani and Genovesi.[10] Auguste Walras' interest in value theory, which has been described at some length by Jaffé,[11] was to show that *rarete* (scarcity) is the *cause of value*. Unfortunately, his definition of *rarete* proved to be rather unrewarding. Leon Walras was determined to continue his father's work but for a long time found it difficult to combine the *rarete* analysis with his own analysis. As he explained in his paper of 1873[12] and the following year in the *Elements*, he had set himself the task of investigating rigorously the advantages claimed for a system of free markets. For this purpose, he developed a model of price determination in interrelated competitive markets. He was able to make good use of Cournot's functional analysis of demand, but after 12 years of painstakingly developing his model, he had not yet managed to work out the relation between utility and demand nor to incorporate the *rarete* idea satisfactorily. He accomplished the desired integration only some time in 1872 when a professor of mechanics, his colleague in Lausanne, showed him how to do it as a maximization problem.[13]

Though Leon Walras had wanted to resolve his father's problems in value theory, his chief concern, as Jaffé has emphasized, 'was the completion of his competitive market model, and not the elaboration of a theory of subjective valuation in consumption'. He had designed a *conceptual machine* for determining equilibrium prices and 'needed a motor to run it'.[14] Nevertheless, it was not altogether a matter of mechanics. Despite his dislike for utilitarianism, his model was then designed to investigate the advantages claimed for free markets on the basis of a criterion (utility maximization) not unlike the utilitarian norm by which a social institution is judged best if it procures the greatest happiness for the greatest numbers. Pareto, of course, was later to steer the Walrasian general equilibrium model away from notions of welfare, happiness, satisfaction and utility towards a more purely mechanistic conception and, ironically, became the father of modern welfare economics in the process of doing so.[15]

c. Pareto

Pareto felt uncomfortable not only about the utilitarian greatest-happiness principle implicit in Walrasian general equilibrium, but also about the notions of pleasure, utility and value themselves. Unlike Jevons, he did not want to explain the nature of wealth and value. In his published correspondence with Croce,[16] he told the latter: 'Let others concern themselves with the *nature*, with the *essence* of *value*. I am interested only in seeing whether I can discover which *regularities* are presented by prices' (italics in the original). In the same letter he also said:

> I look for a theory which may include and present economic facts. For my part, I know only the system of equations of pure economics as being capable of attaining that end, just in the same way that the system of equations of celestial mechanics explains and represents the movement of celestial bodies.

In the earlier of the two letters, Pareto had remarked upon the analogy with mechanics in terms which call to mind the way Jevons had spoken of the 'curious' analogy with the theory of the lever. Referring to the 'equations which determine equilibrium', Pareto said that somebody, possibly the writer, had observed: 'These equations do not seem new to me, I know them well, they are old friends. They are the equations of rational mechanics.'[17]

d. *Mechanics as a hypothesis in the positivistic sense*

When economic theory is formulated in terms of mechanics, it is hardly surprising, not even 'curious', that the equations of economic theory should turn out to be 'old friends' from mechanics. It is not clear whether either Jevons or Pareto saw the matter in this way, or whether they thought they had stumbled upon a profound discovery. Both of them of course lived a long time before the days of Wittgensteinian language-games, Kuhnian paradigms and group-licensed ways of seeing and thinking, which, could they have known them, might have affected their views on this matter. Both Jevons and Pareto, however, may be described as positivists (but not Walras who rejected the idea that his model needed testing and who may best be described as a rationalist[18]). Jevons set out a form of positivism at great length in his work on logic and scientific method, though he did not use the term *positivism*.[19] Pareto made statements such as: 'In experimental sciences the best hypothesis is the one that best relates the greatest number of facts' and 'The demonstration of … theories consist only in seeing whether their consequences fit the facts'. He told Croce that he had been reading Poincare, described himself as 'the most nominalist of nominalists' and had the usual positivistic reverence for a fact conceived as something quite independent of the terms in which one thinks and talks about it. His arguments in support of his views on economic theory can make no sense unless this presupposition is taken into account.[20] Neither Jevons nor Pareto specified the facts against which they thought the new economic theory could be or had been successfully tested, but it is quite clear that both thought the theory could withstand such tests.

Positivism, as explained at some length, is one manifestation of the subjective re-orientation of thought. In the case of Jevons and Pareto, however, the re-orientation affected only the principle on which their commitment to the mechanistic conception rested, i.e. mechanics was not regarded as an expression of ultimate reality but as a valid and useful set of hypotheses. Moreover, both of them (Jevons more explicitly) credited Newton with being one of the founders of a positivistic way of thinking, so that for them there was no change of principle.[21] This is perhaps a questionable view. As related in Chapter 2 above, the positivism of Mach (who wrote mainly in the years between Jevons' and Pareto's expositions of economic theory) began with a critique of Newtonian presuppositions. Be that as it may, Jevons considered it necessary to qualify an unmitigated faith in mechanics.[22]

> The course of nature is regarded as being determined by invariable principles of mechanics which have acted since the world began and will act for evermore.
>
> All that happens whether right or wrong, pleasurable or painful, is but the outcome of the necessary relations of time and space and force.
>
> Such notions I should describe as superficial and erroneous, being derived, as I think, from false views of the nature of scientific inference, and the degree of certainty of the knowledge which we acquire by inductive investigation.

Jevons did not object to the mechanistic conception as such, only to the fact that an elaboration of it seemed 'to be the coming religion'. Pareto also, in his correspondence with Croce, stressed repeatedly that he was quite prepared to drop his mechanistic economics 'if somebody found something better'.[23] His manner of writing, however, suggests that he thought there was very little danger of his ever having to do so.

Jevons and Pareto (more so than Walras) introduced the rigour of the methods of mechanics into economics. The subjective re-orientation of philosophic thought had something to do with this. But it affected only the principle of their commitment to the mechanistic conception. It did not affect their commitment as such nor, above all, did it affect the mechanistic conception itself. If the word *subjective* may still be applied to the price theory which today bears the stamp of Pareto and indirectly of Jevons on it, the subjectivism in question is quite unrelated to the subjective re-orientation of thought. The intellectual descent of that price theory must be traced to the works of Laplace and Newton and to some of the earlier applications of the calculus.

3. The notion of utility

a. Utility has a long history in social theory

The *utility* part of *marginal utility*, as the above discussion has borne out, had antecedents in utilitarianism and in an older tradition of economic analysis. In three long footnotes in the *Grundsätze*, Menger mentioned some 15 writers who tied up the notion of value with the notion of utility or usefulness in one way or another.[24] These so-called value-in-use theories, expounded by various writers in the 19th century and notably by Condillac and Galiani in the 18th, can be traced back via, inter alia, Pufendorf, Davanzati and medieval scholastics to Aristotelian roots.[25] Utilitarianism is equally ancient, though the word itself is of more recent origin. The similarity between utilitarianism and Epicureanism, a philosophy with a long history going back to the 3rd and 4th centuries BC, is generally recognized. J.S. Mill even saw the Socrates of the Platonic dialogues as an exponent of utilitarian philosophy.[26] English classical political economy was strangely cut off from these venerable lines of thought, even though many classical political economists were utilitarians in other spheres of life. The question to be considered in this section is whether, or to what extent, the new economic theories dating back to the 1870s re-established contact between the mainstream of economic analysis and these ancient traditions. The answer to this question will go some way towards showing where in the 20th-century economics an affinity with the subjective re-orientation of philosophic thought is.

b. The interpretation of utilitarianism

Utilitarianism may be regarded as a purely ethical doctrine setting out a norm, but it may also be regarded as a doctrine concerned with what people actually do, namely, seek pleasure and avoid pain. If utilitarianism is understood in the latter sense, many value-in-use theorists may be said to have had a utilitarian outlook. As Schumpeter put it, they were Benthamites by anticipation.[27] This does not mean, however, that they necessarily interpreted utilitarianism as Jevons did. Utilitarianism is susceptible to various interpretations and there are also various popular conceptions of it. The most common idea, perhaps, is that it considers people to be motivated purely by a selfish desire for sensual titillation.[28] When the word *pleasure* is used in this context, we may say (for present purposes) that it is used in its narrow sense, without thereby implying that it is a semantically incorrect sense. Another idea of utilitarianism is very much the opposite of the one just mentioned, implying, in the words of J.S. Mill who criticized the view, 'the rejection, or the neglect, of pleasure in some of its forms; of beauty, of ornament, or of amusement'.[29] The idea is suggested by ordinary discourse in which, for instance, one may be inclined to say that much of 20th-century architecture serves a purpose but gives no pleasure. When the word *utility* is used to indicate something severely practical, we may say that it is used in its narrow sense.

c. Bentham's utilitarianism

Pleasure and *utility* in their narrow senses mutually exclude each other. Since Bentham linked the two words together, he must have used at least one of them in a broader sense. At the beginning of the 'Principles of Morals and Legislation', Bentham stated:[30]

> By utility is meant that property in any object, whereby it tends to produce benefit, advantage, pleasure, good or happiness (all this in the present case comes to the same thing) or (what comes again to the same thing) to prevent the happening of mischief, pain, evil, or unhappiness to the party whose interest is considered

It is plain from the remarks between brackets that Bentham did not insist on fine semantic distinctions. This is also evident in his list of 14 sorts of pleasure, which includes the pleasures of amity, good reputation, power, piety, benevolence and malevolence. In the 'springs of Action' it appears that he replaced the last two with the pleasures of sympathy and antipathy, of which the latter included the pleasure of revenge.[31] Clearly, Bentham did not regard pleasure as a feeling alongside piety, sympathy etc. but rather as a generic term covering all the rest. He used neither *pleasure* nor *utility* in its narrow sense, unless of course one credits Bentham with having seen his fellowman as a rather macabre individual.

Further light was thrown on Bentham's notion of pleasure in the long Introduction to his *Collected Works* of 1845 (Bentham died in 1851), written by the historian J.H. Burton, one of Bentham's two literary executors. Burton, who believed that he 'had a more intimate knowledge than belongs to the majority of general readers, of the nature of Bentham's Works',[32] made the following remarks:[33]

> The terms, *choice* and *preference*, are useful in explaining the meaning of the word *pleasure*, as used by Bentham.
>
> The term nearest to being synonymous with pleasure, is *volition*: what it pleases a man to do, is simply what he wills to do. By considering it for a moment in the light of mere volition, we separate it from the notion of actual enjoyment – that popular acceptation which is most likely to lead us astray. (Italics in the original)

It would be odd to say that pleasure is the same thing as an exercise of the will. It seems that at most a feeling of pleasure could accompany an exercise of the will. However, since Burton spoke of synonymous terms, one should perhaps not approach the matter with the natural attitude and a psychological point of view but look at it instead from a logical point of view as a relation between statements. On that basis, the statement 'A did B voluntarily' would be equivalent to the statements 'It pleased A to do B' and 'A derived pleasure in a broad sense from doing B'. J.S. Mill, who must have been as well qualified to speak on the subject as Burton, pointed out a similar synonymity between the terms *interests* and *wishes*. Commenting on the premise of the 'Bentham school', namely, 'that man's actions are always determined by their interests', he said: 'as the same philosophers, especially Bentham, gave the name of an interest to anything which a person likes, the proposition may be understood to mean only this, that men's actions are always determined by their wishes.'[34]

How does the notion of utility fare if pleasure is understood in a broad sense? Bentham, as shown above, said that utility was the property in an object whereby it tends to produce pleasure. If one has in mind the physical (including chemical) properties of the object, one has to suppose corresponding properties of the human body such that, whenever the object is combined with the body, a glow of pleasure results. Perhaps this is an adequate description of what happens when an apple is eaten or whisky is imbibed, but it is inadequate in the case of the majority of goods said to have utility. The physical properties of a candle or of a knife are not such that a combining of the object with the user's body results in the pleasure of piety or of revenge, but these objects may be instrumental in evoking these *pleasures* understood in the broad utilitarian way. In other words, when one considers utility in terms of reactions between objects and human bodies, one is restricted to the narrow sense of pleasure and one ignores the most important aspect of the notion of using, namely, that an object is used for a purpose, as a means to an end. (Very few of the old value-in-use theorists appear to have ignored this.) The physical properties of objects, or rather those which prospective users believe the objects to have, are of course involved in the notion of the utility of those objects. But however detailed one's knowledge of the physical properties of an object may be, one does not know whether that object has utility unless one also knows the purposes for which it may be used. In the utilitarian way of thinking, that means one has to know what pleasures in the broad sense may be derived from its use.

d. Utility in economics

When guesses are made and general judgements are formed about the purposes for which various types of object (i.e. commodities) are normally used, correspondingly general ideas are arrived at about the utility that various types of objects normally have for people in general. It was utility in this sense that classical political economists had in mind when they mused, like Adam Smith, about the paradox of value and said that it was difficult to find any definite relation between utility and exchange value. The innovation of marginal utility theory was to relate exchange value not to the utility of a type of object for people in general but to the utility which the parties to an exchange ascribe to the actual objects to be exchanged on a particular occasion. In this formulation the connection between utility and the broad sense of pleasure is of course retained. Jevons said that Bentham's definition of utility (see the quotation above) 'perfectly expresses the meaning of the word in Economics, provided that the *will or inclination* of the person *immediately concerned* is taken as the sole criterion, for the time, of what is or is not useful' (italics added).[35]

With the elaboration of the new theory, thoughts of means and ends came to be confined to the entrepreneur acting on behalf of his firm, so that with time the notion of utility to the consumer came to be fused with the notion of pleasure or satisfaction in the broad sense, though this tendency was less evident among the followers of Menger. Von Mises, for instance, defined utility as 'causal relevance for the removal of felt uneasiness'.[36] Jevons distinguished clearly between utility and pleasure but then said: 'Utility must be considered as measured by, or even as actually identical with, the addition made to a person's happiness.'[37] Marshall spoke of the 'total utility of a thing to anyone' as 'the total pleasure or other benefit it yields him'[38] and in present-day elementary textbooks, where the fused notion lives on, *utility* appears to be taken (in line with the natural attitude) to denote a psychological entity perhaps to be measured in utils. When this rather peculiar concept was still in a formative stage and still under general discussion by economists, attempts were made to keep clear of the narrow senses of utility and pleasure or satisfaction by introducing terms such as ophelimity, desirability and wantability measured in wantabs, the latter being Irving Fisher's contribution.[39] Economists therefore found it as difficult as Bentham did to express the broad sense of pleasure.

4. The notions associated with mechanics and utility respectively proved to be incompatible

We have now investigated the notion of utility. Next, we want to see that it did not combine well with the mechanistic conception in economics because of its basis in the broad sense of pleasure and that this in principle proved to be the undoing of the ancient notion of utility in the mainstream of economic theory.

a. The views of Schumpeter and von Mises contrasted

Schumpeter, commenting on the wide meaning attached to the utilitarian term *pleasure*, said it was used with some success to counter 'the allegation that has made human behaviour turn on beefsteaks'. He went on:[40]

> But this success … was more apparent than real. For if we go very far beyond the grossest gratification of the simplest appetites, we come dangerously near to identifying expectation of *pleasure* with all possible motives whatsoever, even with the intentional suffering of pain, and then, of course, the doctrine becomes an empty tautology.

Schumpeter, who despised utilitarianism and took pains to explain that the *unholy alliance* between it and economics did not impair the analytical work of economists,[41] failed to point out that marginal utility theory was based on the same broad sense of pleasure. Any statement about what people do or do not do, even one about the intentional suffering of pain, can be rephrased in terms of more or less utility (as a psychological entity) at the margin, although the result would suffer from some linguistic paucity. Marginal utility theory did not so much provide explanations for what people do as provide a new way of looking at human action. Whether or not one regards this as a defect of the theory depends on what one expects from a theory. Von Mises, for instance, did not regard it as a defect.

In contrast to Schumpeter, von Mises considered the utilitarian *pleasure* to be, if anything, not broad enough. But he put a different complexion on the matter. *Pleasure* was not a more or less comprehensive notion but a formal notion. Writing on Epicureanism, he commented that 'many representatives of this philosophy failed to recognize the purely formal character of the notions pain and pleasure and gave them a material and carnal meaning'.[42] In line with the subjective re-orientation of thought, he regarded pleasure and pain as intelligible forms or terms in which human action is understood. He thought that the formal character of the notion of pleasure had been recognized by what he called progressive utilitarians and he added, by way of illustration, a quotation to the effect that we do not want an object[43] because it is pleasing, but we call an object pleasing because we want it. Furthermore, he said that in the terminology of the branch of science into which, according to him, economics falls, 'the proposition: man's unique aim is to attain happiness, is tautological'.[44] In other words, he took the utilitarian principle to be implicit in the notion of aiming, of volition or simply of human action. As we shall see, he did not regard the implications of the notion of human action as insignificant. In fact, he considered the whole of economic theory to consist of such implications. According to von Mises, economic theory, just as mathematics, deals only in tautologies.[45]

b. Utility theory yields tautologies

The difference between Schumpeter's and von Mises' attitudes may be considered in the terms used in the previous chapter. The formal notion of pleasure and the corresponding notion of utility yield what were called empty (ontological type) explanations in the previous chapter. (See pp 59–68, this volume.) For instance, someone may ask; 'Why did A do B?' and be given the answer: 'Because A derived pleasure from doing B.' If *pleasure* is understood in the narrow sense, the answer may be taken to imply that A is selfish, perhaps neglects his duty and so on. But if *pleasure* is understood in the broad or formal sense, the answer may be rephrased, as we saw above, to read: 'A did B voluntarily.' The questioner may then object that the voluntary nature of A's action was implied in the question and that he is being told that A did B because A did B. Something of such a questioner's objection may be seen in Cairnes' reaction to Jevons' new theory:[46]

> I am wholly unable to conceive how anything amounting to a real explanation can be extracted from the theory we are now considering. What does it really amount to? In my apprehension to this, and no more – that value depends upon utility, and that utility is whatever affects value.

Cairnes invited his readers to suppose that the term x had been used instead of the term *utility*. He continued:

> We might then say that value was determined by x; and the proposition would be precisely as true, and, so far as I can see, as instructive as Mr. Jevons' doctrine. In either case the information conveyed would be that value was determined by the conditions which determine it – an announcement, the importance of which, even though presented under the form of abstruse mathematical symbols, I must own myself unable to discern.

In other words, as a source of ontological-type explanations, Cairnes considered Jevons' theory to be a complete failure.[47]

There can be no doubt that Jevons intended his theory to provide ontological-type explanations. But explanations based on utilitarianism and the marginal utility theory need not be regarded as ontological-type explanations and it appears that von Mises did not so regard them whereas Schumpeter and Cairnes did. As pointed out in the previous chapter, explanations may be of a reflective type indicating the terms in which events or things are understood, or, if one likes, how propositions about them are understood. Such explanations draw implications out of the statements they are meant to explain and in putting forward such explanations one is quite likely to indulge in tautologies. That this does not necessarily make the effort worthless is surely shown by the pursuit of mathematical reasoning throughout the ages. Reflective-type marginal-utility explanations of purchases made are in fact not necessarily tautological. The explanation, 'A bought B because the marginal utility of B per unit of expendable funds was to A at the time greater than that of all other goods', also has implications, such as that A considered other purchases and took into account his current possessions, and these are not necessarily implicit in the bare statement 'A bought B'. However, what is important in the present context is this: A tautology is quite harmless in a reflective-type explanation, but it vitiates an ontological-type explanation.

c. The preoccupation with mechanical notions made utility otiose in economic theory

The notion of utility therefore presented a problem to those steeped in a tradition of thought inspired by the great successes of classical mechanics. Explanations in mechanics need not be regarded as ontological-type explanations either, but the admirers of mechanics apparently did regard them as such. To their way of thinking, there may well be reflective-type explanations, but they are incidental to the main purpose, which is to find a mental representation of a process of determination going on out there somewhere or, if seen in a positivistic light, to find a hypothesis that allows one to proceed as though there were such a known process. The problem with marginal utility theory, from this point of view, was that its form seemed to be suspiciously like the following: Let us define x as that which partly determines y so that we may be in a position to state that y is partly determined by x. The circularity of this argument cannot be removed by calling x utility, satisfaction or pleasure. Whatever interpretation is put on x and y, the form of the argument remains the same. In a scheme of thought in which everything is taken to have determinants, the non-specific statement 'y is partly determined by x' can have any number of interpretations and one may as well leave aside the question of interpretation to concentrate on the form of the statement. As Schumpeter put it when discussing Pareto's contribution to economics: 'We may transcend economics and rise to the conception of a system of undefined "things" that are simply subject to certain restrictions and then try to develop a perfectly general mathematical logic of systems.'[48] If the statement 'y is partly determined by x' describes a part of such a system of undefined things, the logic of the system does not require x to have any specific interpretation, but it does require x to be an *independent* variable.

In his correspondence with Croce, Pareto showed that his thoughts had moved along the lines indicated above. He put forward the proposition: 'So-and-So gets more pleasure out of A than out of B. He therefore chooses A and from this fact certain equations follow.' Pareto observed that if *pleasure* is defined independently of choice (e.g. in the narrow sense), there is no telling whether the proposition is true. If, 'as has been done in reality, one makes that pleasure depend on choice', the proposition is tautological. Either, therefore, it is 'doubtful and causes difficulties' or it is 'useless'. In any case, all that is needed is the independently given 'fact of choice' from which, Pareto presumed, certain equations follow. The notions of pleasure and utility are quite unnecessary.[49]

However, if Pareto's line of reasoning is followed up, it will be seen that the notion of choice is also quite unnecessary. All that is needed for an axiomatic system based on the example of mechanics is a set of conditions from which a conclusion may be derived by stipulated mathematical operations. In the view under discussion, as may be judged from the remarks by Schumpeter quoted above, it is the mathematical logic of the system that really counts and not the garb in which economic questions may come to our attention in everyday life. Words such as *pleasure*, *utility* and *value* may of course be attached to the basic analysis as verbal garnishing – they may even be indispensable as a sop to lesser souls unaccustomed to a scientific mode of thought.

The analogy with mechanics was not foreign to the thinking of classical political economists. J.S. Mill even spoke of a science of human nature which, he thought, could attain the exactness of *Tidology* (the study of tides) though, he admitted, probably not that of astronomy.[50] But the analogy always remained very vague. It was different with the new theories stemming from the 1870s. 'Marginal utility analysis introduced the marginal concept as an instrument of maximization analysis.'[51] The equi-marginal principle made the analogy much more definite; the similarity could now be pinpointed and refined. The new emphasis fell on allocation problems treated with the precision of the axiomatic system of mechanics and economics gained a rigour that political economy had lacked.

But, as is usually the case, the way to gain rigour was to say less. The old notions of pleasure and utility and even of choice and value were incorporated into the new scheme in a way that in effect eliminated them. It might of course have been very different if the pioneers of classical mechanics had thought, for instance, that Halley's Comet enjoyed whizzing past the sun every so many years and planned to do it more often. But that is not how they had conceived the matter and the terms in which most people understand the actions of their fellow-men really have no business in the mechanistic conception. Perhaps this issue has not always been faced uncompromisingly by the architects of economic models who have found so much scope in maxima and minima. But the issue was thought through by the man to whom, more than to anyone else, they owe the style.

Pareto wrote to Croce: 'There are people who think that the new economic theories have been produced to explain value. Far from it! I do my whole course on political economy without using the term value, and I only mention it when I touch on the history of the doctrines.'[52] In Schumpeter's view, marginal utility theory merely served as a ladder by which economists climbed up to the conception of general equilibrium.[53] For that conception only the equi-marginal principle is really essential. As Hutchison has put it very neatly, what was important in marginal utility was the adjective rather than the noun.[54]

5. Menger recognized the incompatibility of mechanics and utility

Hutchison's conclusion, however, does not apply to Menger and that brings us to the final part of the argument of this section. Some years ago, Erich Streissler posed the question: 'To what extent was the Austrian school marginalist?'[55] He came to the conclusion that 'marginalism was not the essence of their endeavour'. 'Marginalism', he said, 'is introduced in the middle of Menger's *Grundsätze*, but it is for this very reason not central to, not the keystone of, this very logical construction'. He pointed out Menger's stress on information transmission, 'interindividual differences in information', uncertainty, market imperfections, divergences between demand and supply prices, price conflict, bargaining processes, cost of exchange, and so on. For Menger, he said, 'everything immediately ramified in some five to ten dimensions' and summed up: 'Menger had incorporated into his founding volume practically all the ideas which make the application of the marginal calculus difficult and hazy.'[56]

The topics which Streissler mentioned are of course to be found in Menger's writings. They are featured prominently by many later Austrian writers and many of them have been discussed more widely by economists of various persuasions in recent years. Yet Streissler's description seems to accord Menger less than his due. It leaves the impression that Menger had profound but unsystematic insights. It seems to endow him with subjectivism as a confounding principle which, though it may enrich a subject matter, also leaves it more perplexing and intractable than it appeared to be before the analysis was started. Such an impression does not do Menger justice. For instance, Streissler spoke of 'Menger's paralysing scepticism',[57] whereas Menger, as we shall see below, had expressly warned his readers to guard against an attitude like that, ascribing it to a mistaken idea of the domain (*Gebiet*) of economics.[58]

Menger firmly believed in a universal principle of causation, as the opening lines of the *Grundsätze* bear out. Why then did the example of mechanics not allure him? It will not do to find the answer merely in an ignorance of that science, or in a 'paralysing scepticism', or perhaps in a peculiar mentality that delighted in floundering in a morass of perplexity. That would leave out of account the ample evidence in Menger's writings that he did not take the marginalist path because his whole outlook was based on a tradition of thought quite different from that in which Jevons, Pareto and many marginalists thought. As we shall see below, the division between different theoretical sciences, to Menger's way of thinking, was based on a broad classification of *Erscheinungsformen.* (Literally, phenomenal forms, i.e. the different forms in which things and events appear.) Theoretical economics had its own set of forms, value and utility being examples. (However, Menger seldom spoke of utility; he referred to the significance attached to the capacity of a thing to satisfy a want.) From this point of view, marginalism would have appeared to be an innovation in economic theory in which the specifically economic forms fade from consideration while the economic theorist concentrates on the forms peculiar to the sciences of mechanics and mathematics – and to Menger's way of thinking, such an innovation would have been downright foolish.

Menger's attitude to mathematics should be seen in the same light. The question has often been discussed. Hayek has remarked that, to the best of his knowledge, Menger nowhere commented on the value of mathematics in economic analysis but that his silence may have been a sign of scepticism. In fact, it has usually been accepted that Menger disliked the use of mathematics in economics. Opinions have differed on the degree of Menger's mathematical competence, though one should perhaps accept as authoritative that of Menger's own son, a mathematician, who did not rate his father very highly in this respect. Hutchison refers to an extract from a letter Menger wrote to Walras (published by Jaffé and not available when Hayek made the above remarks) in which Menger asks rhetorically how one could attain a knowledge of the nature of value, rent, profits, etc. by means of mathematics.[59] On the view put forward here, Menger would not have objected to mathematics as a *Hilfswissenschaft* (auxiliary science, a favourite term with Menger) but rather to its use when it displaced what he regarded as the properly economic phenomenal forma. No doubt, this standpoint does not make sense to one who regards mathematics as a superior language in which anything can be said or at least represented. But that is just the point. Menger saw it differently. To him, mathematics was a rigorous exploration of certain selected phenomenal forms just as economic theory was an exploration of its own phenomenal forms.[60]

Though it is not obvious at first sight, there is a similarity between Menger's standpoint on the matters just discussed and von Mises' standpoint on utilitarianism. We shall return to this similarity. At present it is necessary only to recall that we noticed in von Mises' formal interpretation of utilitarianism, an affinity with the subjective re-orientation of philosophic thought. In Menger's case there is also such an affinity. However, it is not an affinity with any of the later manifestations of the re-orientation, but with its main fountainhead.

6. Menger's Aristotelianism

Menger's deep and abiding interest in the philosophy and method of economics apparently occupied him even before the publication of the *Grundsätze*.[61] His point of view was distinctly Aristotelian.[62] According to Kauder, this was common among Austrian intellectuals in the second half of the 19th century.[63] In the world at large, however, Aristotelianism was as unfashionable in Menger's day as it is today, if not more so. If Menger was aware of the unfashionability of his views – and it seems reasonable to suppose that he must have been – he concealed this completely from his readers. In fact, he spoke of what amounts to Aristotelian essentialism as though no reasonable man could reject it as an intellectual basis.

a. The basis

1. The Grundsätze

Menger outlined his standpoint very briefly in the preface to the *Grundsätze*. He had tried, he said, to resolve the complicated phenomena of economic life into the simplest elements still open to certain observation; then, focusing attention on these, to investigate how complicated economic phenomena develop according to law out of their elements.[64] This is of course not a very usual way of stating an Aristotelian position, though it was not foreign even to Aristotle himself. It is interesting that the very first footnote reference in the *Grundsätze* is to Aristotle's *Politics* I.3 and that in *Politics* I.2 (i.e. one or two pages earlier in a book of normal size) there appears the sentence: 'As in other departments of science, so in politics, the compound should always be resolved into the simple elements or least parts of the whole.'[65] In his later writings, Menger did occasionally use typically Aristotelian expressions. For instance, when distinguishing between cognition and understanding (*Erkenntniss* and *Verständniss*) in the *Untersuchungen*, he said that we understand phenomena when we (re)cognize the grounds of their being and their so-being (*den Grund ihres Seins und ihres So-Seins*).[66] The expressions *being* and *so-being* are a legacy of Aristotle's distinction between primary substance and substance as an object of thought, i.e. of the distinction which accommodated the fact that everything is known in terms of universals (see p 24, this volume).

Menger used a variety of terms for the so-being of economic phenomena. In the place where he spoke of so-being, he used the term as a synonym for *eigenthümliche Beschaffenheit* (characteristic constitution) but more commonly he used the word *Wesen*. This word is often translated as *essence*, especially in an Aristotelian context, and since the etymology of both words seems to indicate an original meaning of *being*, it is no doubt a correct translation. But *Wesen* in German usage does not have the connotation of something abstruse that *essence* has in English usage so that the word *nature* perhaps conveys the meaning more accurately. Thus, the title of the first section of the first chapter of the *Grundsätze*, namely, '*Ueber das Wesen der Güter*', simply means 'On the Nature of Goods'. When Menger wanted to reduce the nature of something to its elements (as he always set out to do), he spoke of *Erscheinungsformen* (phenomenal forms) and, when these were recurrent, of *Typen* (types) to which he added the word *strenge* (strict) when he wanted to indicate that they were the simplest elements, the simplest, as he sometimes suggested, that could be thought.[67] The study of strict types in economics – the morphology of economic phenomena, as he eventually called it[68] – was of fundamental importance to Menger, though it was never an end in itself. It systematizes what is in any case done in everyday life. Types are identified through a recognition of similarity (though this is not quite how Menger put it). Without a knowledge of types, one could not have a mental grasp of the world of experience and without a knowledge of the typical relations between phenomenal forms one would forgo not only a deeper understanding of reality but also the possibility of going beyond immediate experience, i.e. of foresight and a control over things.[69]

All this is consistent with an Aristotelian outlook, but it could also be called an unusually elaborate articulation of a common-sense approach to a subject. As examples of the phenomenal forms of economics, Menger cited purchase, money, supply and demand, price, capital and interest.[70] An author who singles out one of these for his attention and begins by discussing the nature of his subject matter and then proceeds to reduce it to its elements, i.e. to analyse it, would not be doing anything out of the ordinary. An author who really understands the fact-hypothesis dualism and takes it seriously would of course avoid any discussion of the nature or essence of things. But a thoroughgoing positivism is probably less common even today than a basic essentialism. Menger, it must be remembered, did not claim to have put forward an unusual viewpoint; in fact, he more or less claimed the opposite. Nevertheless, Menger's Aristotelianism has certain features which differ from some commonly held views and in the present context these features are the most interesting.

2. Exact laws

In the preface to the *Grundsätze*, Menger raised two related issues which he was to enlarge upon in his later writings and on which some later members of the Austrian school were to differ somewhat from him. He claimed that the method of research he had used (i.e. reduction to elements, etc.) was the same as that which had been used to good effect in the natural sciences (in which he included the physical sciences).[71] He frequently drew parallels, especially in his later writings, between chemistry and physics on the one hand and economic theory on the other. However, Menger can easily be misunderstood on this point. The common method he had in mind extended no further than the reduction to elements or types and to the typical relations between types.

He was to clarify his ideas on this, at least in print, only much later. As he eventually explained it, he saw two criteria for the division between the various sciences. There was division by subject matter, i.e. by kinds of phenomenal form or simply the nature of the things studied. By this criterion there was of course a difference between economics and the natural sciences. But there was also division according to the direction of cognitive endeavour (*Richtung des Erkenntnisstrebens*) or the way of seeing (*Betrachtungsweise*).[72] By this criterion there could be many types of sciences but the cognitive direction and method (reduction to types and typical relations) that Menger was eager to promote led to what he called *exact* science and, in this sense, there could be and were exact natural sciences and an exact science of economics. Menger tended to attach more importance to the second criterion. In one place in the *Untersuchungen*, for instance, he remarked that the circumstance that research in the field of human action proceeds from the presupposition of a definite intention (*Willensrichtung*) on the part of the human agent, was a characteristic (*Eigenthümlichkeit*) peculiar to the social sciences and did not establish an essential difference between the exact natural and social sciences. The exact natural sciences also had characteristics peculiar to themselves.[73] The subject matter, it appears, merely introduced peculiarities into a science.

Menger must already have had something of all this in mind when he wrote the *Grundsätze*, for immediately after pointing out that his method of research was the same as that which had led to the great successes of the natural sciences, he said that every method acquired peculiarities according to the sphere of science in which it is applied and insisted that for this reason there could be no question of using the natural-science approach in economics. The attempts which, according to him, had been made to carry the peculiarities of the method of the natural sciences over to economics had led, he said, to an empty playing with superficial analogies between the phenomena of economic life and of nature.[74] He did acknowledge that one should not lose sight of the ultimate goal of establishing the connection between all the sciences and the unity of their highest principles, but he thought that this goal could not be attained until the individual sciences had been thoroughly explored.[75] We have already seen that Schumpeter was to credit Pareto with having advanced towards such a goal, i.e. with having transcended economics to the conception of a system of undefined things subject to certain restrictions. One may be certain that Menger would not have agreed with Schumpeter, that he would not have put the formalizing inspired by axiomatic mechanics into the category of establishing the highest principles of all the sciences but rather into the category of playing with superficial analogies.

To arrive at Menger's conception of exact science, one has to set aside all considerations of a specific subject matter and for this purpose *system*, *restrictions*, etc., must be counted as phenomenal forms belonging to certain subject matters. The pursuit of exact science is simply a certain way of treating any subject matter whatever it may be, a certain direction of cognitive endeavour. It may seem that such a remote notion could hardly have mattered when Menger turned to his specifically economic analysis. But this is not so. It did influence his approach to certain problems in economics which even today are far from resolved. This approach already came into play in the case of the second issue Menger raised in the preface to the *Grundsätze*. He warned his readers not to heed the opinion of those who, from a consideration of human free will, denied the regularity of economic phenomena (*Gesetzmässigkeit*, thus regularity in the sense of conformity to law or principle). On that basis, he thought, the possibility of economics as an exact science was negated altogether and that of course he was intent to deny.[76] He conceded that free will was a good reason for not expecting the full regularity of human action itself and in a later work he said that he did not wish to deny free will as a practical category.[77] But he considered the objection to be irrelevant. Free will did not preclude an exact science of economics. Menger was later to treat the question of error and ignorance on the part of the economic agent in an analogous way and this gives us a clue to what he might have said in the present-day discussions among economists on the subject of uncertainty. His approach to the question of free will is therefore of some interest. Unfortunately, the argument which Menger put forward in 1871 in support of his contention was rather obscure. Among other things, he said that economic theory is related to economic activity as chemistry is related to the activities of the practising chemist.[78] For a clarification of his views in this regard, one must turn to the *Untersuchungen* and his even later writings.

b. Free will, error, ignorance

1. The Untersuchungen

On Menger's own admission, the *Untersuchungen* was a polemical work.[79] It was directed against the historical school of economists which dominated academic economics in Germany during Menger's lifetime, and especially against Schmoller, its leading figure at the time. Menger charged the historical economists with not being able to distinguish between a historical and a theoretical understanding of economic phenomena. The book begins with a discussion of two main directions of research or cognitive endeavour. In one, attention is focused on concrete phenomena in space and time and on relations between them while in the other it is focused on recurrent phenomenal forms and on typical relations between them. The former concentrates on the *individual*, the latter on the *general* aspects of phenomena. By *individual*, Menger explained, he did not mean the opposite of *collective*, for which in the early parts of the book he preferred the word *singular*. The mark of the *individual* was a definite spatial and temporal setting. (Menger's distinction between individual and general is of course similar to the distinction made in the previous chapter, but they are not identical.)

When the two main directions of research are brought to bear on economic phenomena, Menger argued, quite distinct disciplines arise. Concentration on individual aspects leads to historical and statistical studies. Concentration on general aspects leads to theoretical economics which includes but is not co-extensive with the exact science of economics. In addition, Menger recognized practical sciences of economics (he gave the example of finance) which were not concerned with cognition at all but with making practical suggestions for achieving certain aims and which drew on other disciplines for this purpose. He conceded a useful role to all these subdivisions of economics, provided that their separate identities were respected, but by implication he put economic theory into a pivotal position because the other sides of economics had to use it as an auxiliary science.[80] Menger had formed the impression that the German historical economists were going about their task as though they were combining the historical, statistical, theoretical and practical sides of economics into one comprehensive science. The main purpose of the *Untersuchungen* was to show that such a science would be a monstrosity.

2. *Empirical and exact laws*

However, to understand Menger's attitude towards free will, error and ignorance on the part of economic agents, one has to follow up a subsidiary issue in the development of his argument. He introduced a subdivision of the theoretical direction of research which apparently had not occurred to him when he wrote the *Grundsätze*. He now spoke of the *realistic-empirical* and *exact* directions of theoretical research.[81] All theory, as before, was concerned with the general, more specifically with typical relations or, in the phrase which Menger never tired of using, with the regularities in the coexistence and sequence of phenomena.[82] But the particular with which the general was contrasted was now referred to constantly as full *empirical actuality*[83] – an expression which was used in the sense of actuality as it is experienced and which thus corresponds quite closely to the expression *grassroot factual judgement* used in the previous chapter. In Menger's terminology, the realistic-empirical direction of theoretical research extracted from phenomena in their full empirical actuality certain empirical (as opposed to exact) laws about the coexistence and sequence of phenomena and it is therefore nothing other than what is normally called induction. Menger himself occasionally spoke of induction in this context, such as when he praised Aristotle for having already denied the strictly scientific character of induction.[84] Menger conceded a useful role to empirical laws in economics, but they were no more than rough-and-ready rules of thumb to which there were countless exceptions and for which there was no guarantee of their continued usefulness.[85] In contrast, the exact laws derived by the exact direction of theoretical research did not have these shortcomings.

Just how the exact direction of research could arrive at regularities in the coexistence and sequence of phenomena, if not by induction, is not clear. It is the kind of question which all who have professed an essentialism seem to have found difficult, not least Aristotle himself.[86] The principle which Menger enunciated in this regard is hardly enlightening and it plays no further role in his subsequent discussion.[87] His idea of exact laws is perhaps better understood from the contexts in which it appears, though an entirely clear picture does not emerge either. He indicated, for instance, that exact science was not to be understood as conceptual analysis nor exact laws as deductions from a priori axioms.[88] If they were taken as such, he thought, the exact direction of research would have a methodologically subordinate position in so far as exact laws would have to be corrected by empirical laws whenever a contradiction arose between these types of laws.[89] As Menger saw it, this was far from the case and yet much of what he said about the exact direction of research did suggest conceptual analysis. He said, for instance, that exact laws must hold without exception because, according to the laws of our thought, they simply cannot be thought to have exceptions.[90] Again, he said that exact research does not arrive at exact laws of the actual phenomena of economic life but at laws of economizing (*Wirtschaftlichkeit*).[91] In a book review written in 1887, Menger upbraided one of the authors for dealing only with empirical laws. Mentioning laws of economizing again, Menger described these as laws of or rational economic means-ends relations (*Gesetze der rationalen ökonomischen Zweckbeziehungen*).[92] Little wonder, then, that von Mises was later to revive Menger's exact laws in the form of a priori concepts and propositions.

The concepts of the phenomenal form and typical relation are fundamental in Menger's analysis. They are clearly examples of what was termed the intelligible form in the previous chapter. As we saw there, the intelligible form has appeared in many guises, some more inclusive than others, from the universal and essence in the sense of so-being to innate ideas, a priori categories, terms and so on. Moreover, there has been a tendency for successive interpretations to reflect the increasing self-awareness of the knower. In earlier times at least, these interpretations were often presented as insights into what Aristotle had really meant; not without some justification, since Aristotle approached almost every subject from a logical, sometimes even linguistic, point of view. Ockham, writing some five and a half centuries before Menger, had already interpreted the Aristotelian universal as a purely mental concept represented by the spoken and written word and in his mature years had identified the universal with the act of intellection itself.[93] Ockham is often credited with having founded nominalism, but the *via moderna* to which he contributed was more broadly an early manifestation of what we have called the subjective re-orientation of thought. Since Menger's analysis fitted into a tradition of thought in which many excellent minds had found a bewildering variety of interpretations of the intelligible form, it is perhaps not surprising that his exact laws should evince some ambiguity.

However, if one does not probe too deeply, it is not too difficult to imagine what Menger had in mind when he distinguished between exact and empirical laws. The kind of analysis he had presented in the *Grundsätze* was somehow different from a statement of a connection that is empirical in the original meaning of the word – such as when someone claims that garlic is good for gout. The relation between value and the significance people attach to the capacity of a good to satisfy a want is somehow necessary and intelligible whereas the claimed connection between garlic and gout is not. There is nothing in the layman's understanding of what garlic is and what gout is that makes it intelligible to him what, if any, the connection between them is. If there is a connection, it is just a dumb fact whereas value theory speaks for itself.

Menger often used the word *Bedingung* in the context of his exact laws. This word has a most usual meaning of a condition or a stipulation, but Menger used it in the sense of a rational requirement. One may conjecture that this usage has its ultimate origin in such of Aristotle's remarks as that 'to know a thing's nature is to know the reason why it is'.[94] We have already seen that the Aristotelian essence could be stated in terms of causes, where the current English word *cause* is rather a misleading translation of the original, which is perhaps rendered better by the German word *Bedingung*. Thus, we understand the nature of prices better when we realize that valuations are a *Bedingung*, a rational requirement for a price. If, for a moment, we do not think of demand and supply as non-specific functions into which specific values of the parameters and variables may be inserted to represent a specified situation, then what does modern price theory tell us but the rational requirements for prices and price changes, the exact laws linking the phenomenal forms of a price and a change in price to other phenomenal forms. That, at any rate, appears to be how Menger would have seen it, and he was convinced that there is a clear-cut distinction between exact and empirical laws.

3. Menger's attack on the historical school

Having expounded the distinction, Menger proceeded to use it as a stick with which to beat the historical economists. He charged them with being confused about (what he called) the most elementary principles of scientific method.[95] They had rejected theoretical economics, as Menger understood it, on false grounds in so far as they had insisted on judging it by a criterion appropriate only for empirical laws, namely, that it should be in accordance with full empirical actuality. Empirical laws, being gleaned from full empirical actuality, must continue to be confirmed by it. They always have exceptions but, if they have many exceptions, they become worthless and simply fall away. Not so, exact laws. Tested in this way, they will always seem inadequate. An exact law provides a theoretical understanding of only one aspect of actual phenomena, and neither can be nor need be verified by full empirical actuality. One need not test the principle of economizing against what people actually do. Menger likened such a test to an attempt by a mathematician to verify geometry by measuring actual objects.[96] A few years earlier, Walras had also alluded to economics pursued *more geometrico* – in the manner of geometry.[97]

It was, however, in bringing this line of reasoning to bear on the historical school's objections to the prominence given in economic theory to self-interest on the part of the economic agent that Menger showed his attitude towards the question of free will, error and ignorance most clearly. He outlined the historical school's strictures with regard to self-interest, which in this context was understood in a narrow sense (in the terminology of the previous section). Mankind acts upon countless, to some extent incompatible, motives and this precludes the strict regularity of human action. Political economy in the Smith tradition makes self-interest into a fundamental axiom. Only by thus falsely regarding man as acting upon only one motive can the factor of free will (*Willkür*) be excluded and economic laws without a temporal and spatial context be conceived.[98] Menger agreed that the multiplicity of motives precludes the strict regularity of human action. He then added, as it appears in the English translation:[99]

> But there is another factor, equally important, that does the same thing. I mean *error*, a factor which surely can be separated still less from human action than custom, public spirit, feeling for justice, and love of one's fellow man can be separated from the economy. Our historians are too lenient towards their scholarly opponents. The presupposition of a strict regularity in economic phenomena … includes not only the dogma of ever-constant self-interest, but also the dogma of the *infallibility* and *omniscience* of men in economic matters.

Commenting on this passage, Hutchison has remarked:[100]

> In emphasizing this assumption of correct knowledge, and the exclusion of ignorance and error, Menger was taking the first step towards the

> opening up of the analysis of expectations. … But in doing so he was admitting that his *exact* laws had a very limited applicability to actual economic behaviour and its prediction.

And again: 'Menger himself seemed to admit the seriously limiting nature of the abstraction involved here when he implied that the historical critics had been too forbearing regarding the limitations of economic theory.'

Hutchison is quite correct, as we shall see, that Menger's treatment of error and ignorance has had at least something to do with opening up the analysis of expectations within the Austrian school. No doubt, he is correct also in implying that Menger's exact laws had severe limitations: one could go further and say that they probably would not stand up to closer scrutiny. But Hutchison is surely wrong in suggesting that Menger was admitting anything, that he was conceding a point to the historical school. The rather heavy-handed sarcasm which Menger reserved for members of the historical school is quite unmistakable in the pages following the passage quoted above. It makes it quite plain that he was conceding nothing to them. Menger went on to say what astonishing folly the greatest minds of all nations had evinced for thousands of years in striving after strict social theories and under what lamentable misconceptions mankind would still labour if the historical school of German economists had not opened its eyes. Moreover, since the same misconceptions were found in other fields of science, also chemistry, physics, mechanics and mathematics were untenable, worthless and in need of reform, deplorable confusions of human thinking. And no one had suspected it until the historical school had opened our eyes to it, partly with the instinct of genius, not fully aware of the epoch-making upheavals in the sphere of exact research. Truly, our historical economists can take pride in these their achievements![101] After carrying on in this vein for more than two pages, Menger even announced to his readers that he was returning to the serious side of the matter.

4. Signals and noise

Nor was this the only place where Menger discussed this issue. In Appendix V, he again said that free will precludes a strict regularity in the phenomena of human action. Similarly in Appendix VI, where he linked free will with error, and in the text, where he linked error with ignorance.[102] The word he used for the latter (*Unkenntnis*) means ignorance in the sense purely of a lack of knowledge, in which sense one may be ignorant of the future. Error he also appeared sometimes to understand in the sense of uncertainty. Paragraph 4 of the first chapter of the *Grundsätze*, headed 'Time-Error', deals with what we would now call time and uncertainty. The heading was apparently changed to read 'Uncertainty' in the revised edition of 1923.[103]

What is significant for present purposes is that Menger was not in the least perturbed by a whole range of questions which, seen under various aspects, troubled the historical economists and many others since. Human action in its full empirical actuality, i.e. what people manifestly do before one's eyes, was simply not explicable by exact laws and yet Menger left no doubt that he regarded research into exact laws as the economic theorist's highest calling.

It is hard to understand this attitude unless one takes into account that Menger's is a voice from another era. One has to be able to see in Menger's 'full empirical actuality' something of the 'perceptible flux' of antiquity. In his Aristotelianism, things and events have *Bedingungen*, rational requirements, which define their essential natures, at least from a certain point of view. But things and events also have accidental features and Aristotle's distinction between essential and accidental features was not without antecedents. Behind it lies ultimately the Heracletian question and the quaint Parmenidean discovery of logic (see Chapter 2, Section 2).

There was a time when the attitude that Menger brought to bear on economics appears to have been quite common and it has lived on in sundry ways until quite recent times. (The economic man is not the whole man. An empirical measure is one guided by mere experience and not by scientific knowledge.[104]) However, over the past 400 years this attitude seems largely to have been displaced. One may speculate that the development of celestial mechanics out of astronomy has had something to do with this. Observation of celestial bodies, at least over the distances involved, presented no flux, no intractable welter of experience, but only law-like movement. One may speculate further that the conviction that all else may be reduced to such law-like behaviour arose out of the awe which these scientific successes inspired. At any rate, the great influence of this conviction, especially in modern economics, can hardly be denied. The old attitude, admittedly, appears to have been revived in the signal-noise metaphor current among physical scientists as a *facon de parler*. (One pictures oneself intently listening to a radio, trying to pick out a signal from a background of atmospherics.) In terms of this metaphor, Menger was listening for signals amidst the noise. But economic theory has more commonly been pursued on the presupposition (to continue the metaphor) that there are only signals, albeit so cacophonously intermingled that the listener is left with the impression of more noise.

The importance of this difference in presupposition should not be underrated. For example, it seems natural now to say that Menger's above-quoted remarks about error and ignorance refer to the *assumptions* of infallibility and omniscience in economic theory. Menger did say that highly unlikely conditions would have to be met for full empirical actuality ever to be in accordance with exact laws. Nevertheless, the word *assumption* is inappropriate. Menger was looking for exact laws; he was not constructing models, nor did he think that the political economists of what he called the Smith school had constructed models.[105] In the way the matter is commonly thought of nowadays, the seething welter of economic life is approximated by models reinforced with a battery of simplifying and ceteris-paribus assumptions. With the belief in universal law-like behaviour, ever closer approximations to economic life must be considered attainable, provided that the increasing complexity can be handled, by the judicious addition of more parameters or assumptions otherwise incorporated. (Money illusion, risk aversion, rational expectations, etc.) Alternatively, on a positivistic or Popperian view, ever more of the continual stream of facts passing by before our eyes must be accounted for as unfalsified hypotheses accumulate. Whichever the view, as the net is spread wider there must eventually be included, or perhaps accounted for, not only creativity but also error and ignorance, not to mention frivolity, forgetfulness, stupidity and so on. Since this appears to be a daunting task, the fullness of human conduct becomes something of an embarrassment. In their private lives, economic-model builders presumably understand the conduct of their fellow-men in its familiar fullness, but, apparently by mutual and tacit agreement, it seems they do not let the embarrassing everyday viewpoint intrude too often upon their professional discussions. Menger, however, was not in that position. Whatever he may have said about causality at the beginning of the *Grundsätze*, the rest of the book bears out that the fullness of human conduct was not in the least an embarrassment to him. He spoke of economic life as he was acquainted with it. It was the obvious and natural thing to do and not a matter of a 'paralysing scepticism'. After all, he was not making assumptions nor building models. He probably did not even consider the possibility of that procedure sufficiently to be sceptical about it.

c. Menger and the modern Austrians

1. The similarity

Whatever else Menger may have left to later economists, he bequeathed to the Menger-Mises line of the Austrian school a readiness to see human conduct in its familiar everyday fullness not only as laymen but also as economists. Indeed, one could almost say that this readiness has become the hallmark of modern Austrian economics. Writing at a time when mainstream neo-classical economics is conspicuously threadbare in this regard, the modern Austrians combine an emphasis on human action with a critique of neo-classical economics. As Lachmann has put it, the Austrians distrust 'all those formulations of economic experience that do not have an identifiable source in the mind of an economic actor'.[106] The approach they have adopted they have called *subjectivism*. Kirzner has described it as follows: 'The foundation of subjectivism in the analysis of social phenomena consists, of course, in the insight that these phenomena are generated by deliberate, purposeful human action.'[107] The modern Austrians make the point that one cannot understand economic life unless one takes into account the intentions of those who contribute to it, their plans for achieving their objectives and hence also their knowledge and expectations. Reflection on ordinary experience tells us that these are factors we do in fact take into account when we try to understand the conduct of others.

We have seen that Menger said that research in the field of human action proceeds from the presupposition of a definite intention on the part of the person who acts, that he said this in the context of phenomenal forms peculiar to certain sciences and that he also spoke of playing with superficial analogies. We may therefore conjecture that he might have expressed the attitude of his intellectual heirs in the following terms: It is very important to point out the phenomenal forms in which everyday economic questions arise, i.e. to identify the nature of the subject matter of economics correctly, especially if one is convinced that one's colleagues in another school of thought are engaged in an empty play with superficial analogies between the phenomena of economic life and the phenomena of nature seen under the aspect of mechanics. But what of modern Austrian analysis apart from the identification of the economic subject matter? Menger might well have regarded the analysis of the role of plans, expectations and knowledge in choice as an extension of his own analysis of 1871. He had regarded error and ignorance as irregularities in economic life. They had been part of the noisy background, but now are being recognized as signals after all. In one place, Menger said the most urgent task of the economist was to clear away misconceptions and to fill in the gaps in the system of exact laws.[108]

2. The contrast

These are not the terms in which modern Austrians see themselves, but perhaps they would not object too strongly to being described in them. However, if the spirit of Menger were given the liberty to continue in the same vein, the contrast of paradigms would soon begin to jar and eventually there would be a clash. Let us give Menger the opportunity to speculate about the goal of exact research, what it may all lead to. At the place where he spoke of filling gaps, Menger raised the possibility that full empirical actuality may be understood in its entirety by the exact laws of a great multiplicity of sciences. It is the rationalists' dream that ultimately everything can be shown to be such that it could not be thought to be otherwise. Menger did not take the possibility very seriously. He merely said that if mankind should ever achieve a universal exact understanding of social phenomena, it would be along such lines. He thought that the development of the many other sciences should be accompanied by the best wishes of economists, who nevertheless should concentrate on filling in gaps in exact economic theory.[109] In other words, he thought that one should pursue the rationalists' dream as far as one can, but that, in the nature of things, it is unattainable. One has, therefore, always to ask oneself whether one is dealing with plans, expectations and knowledge in full empirical actuality or as phenomenal forms and never should one fall so low as to confuse the two, like the historical economists. Now, Mises was to dwell at length on the difference between theory and history and modern Austrians do of course distinguish between theory and historical as well as present fact. No doubt, they also know that this distinction has nothing to do with the subjective nature of the economic subject matter. Yet it is just in this area, which on the evidence of his writings Menger considered his strongest, that his influence at present is weakest.

From a modern Austrian viewpoint, there is an unbridgeable gulf between the social and natural sciences because the former do, and the latter do not, deal with objects to which some human mind has attached a meaning. For Menger, as we have seen, this difference was merely a matter of a peculiarity of subject matter. It should be recalled that he recognized two criteria for the division between sciences, namely, subject matter and direction of cognitive endeavour. All that he was so eager to explain about theory followed from the second criterion and applied alike to the theories of all sciences whatever their subject matter may be. He stressed this point in the text of the *Untersuchungen* and then devoted an entire Appendix to it.[110] Menger was preoccupied with the distinction between the general and the particular and saw it in the form of a difference in the direction of cognitive endeavour. His implicit subjectivism was suited to this whereas the subjectivism of the modern Austrians is not. In the end, therefore, it appears that the subjectivism of the latter is not the same as Menger's. When the modern Austrians insist on 'the subjective character of the data of the social sciences'[111] they are referring to the subjectivity of the objects of the subject's attention. Menger's idea of the forms in which things appear (phenomenal forms) and his insistence on different directions of cognitive endeavour were based on the subjectivity of the attending subject himself.

d. Menger did not solve the problem relating theory to fact

We have now considered three approaches to demand and value theory, namely, marginalism on the analogy of mechanics, very cursorily so far, the subjectivism of subjective objects, and the subjectivism of the subject. They will be recurring themes hereafter. Of course, these approaches are seldom found in isolation. Not very many marginalists appear to have seen the logical implications of their approach as clearly as Pareto. Alongside the subjectivism of subjective objects, modern Austrian economics also has derivatives of Menger's exact laws (with their basis in the subjectivism of the subject) such as the pure logic of choice.

Of the three approaches, Menger's subjectivism of the subject has the clearest affinity with the re-orientation from the natural to the reflective attitude. On the argument of the previous chapter, this approach therefore should in principle bear the seed of a resolution of the problem of theory and fact in economics. But it should also be clear from that chapter that Menger's treatment of the distinction between general and particular had a serious shortcoming. In his discussion of the *Untersuchungen*, Hutchison also remarked upon this shortcoming: 'But rather than simply to distinguish, Menger seems sometimes to suggest the separation of the two types of knowledge, whereas surely the more important methodological questions relate rather to how general and individual propositions are in different ways *combined*' (italics in the original).[112]

We shall have occasion to notice in the remainder of this chapter that Mises and other Austrians still had difficulties over these questions. It is evident once more that some scheme for understanding the relation between general and particular is needed. Such a scheme could show, or so it will be argued in various places, that many of the conclusions derived from the subjectivism of subjective objects may also be derived from the subjectivism of the subject. In fact, the contention will be that they should be so derived.

Appendix 1: 20th-century strands of Austrian economics

The Austrian school was never particularly homogeneous, and this certainly appears to be true also of the modern Austrians who have found recruits on the one hand among those who were attracted by its strong free-market stance and on the other among those who simply liked its message of re-incorporating into economic theory human beings as we know them from our everyday experience. Under the circumstances, one cannot expect to find strong analytical principles that are shared by all.

Where did Lachmann, for instance, stand within this great variety? Here one must adopt the manner of an Aristotle or a Marshall and say that in one way his subjectivism was not that of the original Austrians, or at least of Menger and Mises, but that in another way he was very close to Menger's original conception of how to do economics. What follows is an attempt to explain this contention. Lachmann's 1943 *Economica* article on expectations elicited a response from von Mises later that year.[113] Mises quoted from his own works to show that he had not neglected a point Lachmann had made about 'elastic expectations', though a reader of von Mises would have had to surmise from the context that he was indeed referring to expectations. Mises also objected to the word *elastic* as 'an inadequate and misleading mechanical metaphor' – a little unfairly, since Lachmann's aversion to such things was already apparent in his article. (In his obituary article for Schumpeter in 1950, Lachmann pointedly made a remark about economic theorists – not Schumpeter – 'whose nursery had been equipped with such up-to-date mechanical devices as multipliers, accelerators, and elastic expectations'.[114]) Lachmann said on a number of occasions that Mises' comment had given him the impression that expectations were to Mises a novelty. But there was more to it. Mises had apparently made a disparaging remark to him in private to the effect that the question of expectations was not of much relevance. Though it was not in his nature to show such things, a hint of dormant resentment was still noticeable all those years later.

In view of this, the article Lachmann published in the *Journal of Economic Literature* in 1976 drawing parallels between Mises and Shackle was a little surprising.[115] Though he made the point that Shackle 'extended the scope of subjectivism from tastes to expectations' and that Mises and Hayek had failed to grasp the opportunity of doing so in the 1930s, the similarities he did list were somewhat forced – perhaps not when held up against a caricature of the neo-classical economist, but superficial otherwise. Lachmann did regard it as strategic to combine Shackle's supporters and the new Austrians into a united force (he tended always to see the academic field as a battleground) and of course he had come to admire Shackle immensely. But whether this was the real reason for the article, he was not prepared to confirm or to deny.

In 1978 Lachmann was asked to write the Foreword for the re-issue of Mises' *Epistemological Problems of Economics* and in the course of preparing it he came across some remarks by Mises to the effect that the ultimate ends of action must be taken as given. Such a proposition was not acceptable to the 'subjectivism of active minds' (as Lachmann called his and Shackle's position) and he was visibly upset. For some months he would come up with remarks such as: 'But, of course, Mises tells us that our ends are given to us each morning on a silver platter.' (In this case it was a bit unfair to Mises who clearly had meant that the economic analyst takes other people's purposes as given; and in any case he was merely saying that his praxeology was *wertfrei*.) In his paper, 'Ludwig von Mises and the Extension of Subjectivism', published in 1982, Lachmann finally acknowledged the differences between Mises' and his own subjectivism.[116] He also put his finger on where the difference lies.

Mises, as Menger in a different way before him, was interested in necessary connections in economics and he couched his analysis of these in terms of the a priori. This is of course a most troublesome notion, but Mises made it fairly plain what he meant by it. We cannot speak of success and failure or of profit and loss without having in the back of our minds the notion of means and ends, of purposeful action. We cannot in a concrete situation speak of economizing unless we have classified some things as means and others, possibly imagined states, as ends. Means necessarily imply ends. Mises would have said of this statement 'that the negation of what it asserts is unthinkable'.[117] He might well have conceded that we cannot think of decision and choice without also thinking of the expectations on which they are based. But it is not at all this aspect of expectations which is of the greatest consequence in Lachmann's and Shackle's conception of the role of expectations in economic affairs.

Shackle and Lachmann focused attention on the creative acts of the mind, on the *Protean* nature of thought and the Kaleidic social world to which it leads (both Shackle's terms and great favourites with Lachmann). We may know something of the past, but we can only imagine the future and in making and revising our plans we have to interpret the past. Here, then, the creative acts of the mind enter the arena, for the same news item, the same stock exchange quotation, may be interpreted differently in each individual mind and lead to entirely different pictures of the future. But this subjectivism of interpretation, as Lachmann used to call it, does not pass Mises' test of the unthinkable negation. (Nor, as Lachmann pointed out, does von Mises' theory of the credit cycle.) We may certainly imagine a society so regimented in thought that everyone in it always comes to the same conclusions. It is just that we know that by and large we do not live in such a world. Sometimes Lachmann expressed himself in Misesian terms:

> We are able to imagine a world in which tastes do not change but unable to imagine one in which knowledge does not spread from some minds to others. Even continuity of ends does not entail an invariant means-end pattern; men would still be eager to make better use of the means at their disposal. *Time* and *Knowledge* belong together. (Italics in the original)[118]

It is of course much easier to say what we can imagine than what we cannot imagine. One is reminded of Wittgenstein's remark about the man who said it was self-evident to him that it never rains at the North Pole. But be that as it may. Lachmann's and Shackle's subjectivism is all about the interplay of time and the volatile, idiosyncratic, unpredictable thinking of our fellow human beings. On this they were entirely at one. In a letter dated 14 July 1970, Shackle wrote to Lachmann:

> The passages which I quoted from your *Metroeconomica* article state the essential core of the matter of time and knowledge with unsurpassably brilliant concision. I read them then, I read them now, with the profoundest admiration. Our ways have lain apart, but we have converged in thought to a degree which gives me immense consolation and encouragement in the face of the inimical climate of our time to our subjectivist views.[119]

The subjectivism of active minds or of the Kaleidic society has its antecedents in the works of Keynes, especially in the famous *Quarterly Journal of Economics* article of 1937, as also more faintly in those of Marshall and possibly of Lindahl and Myrdal. But it is not the subjectivism of Menger and Mises for whom subjectivism was much more a matter of the logical nature of economics. Lachmann admired Keynes to a degree that is unusual in Austrian circles[120] and in the last few years of his life the idea of promoting closer co-operation between post-Keynesians and Austrians was very close to his heart. Keynes' brand of subjectivism, especially as extended and elaborated by Shackle, was just what economics needed.

It has of course been quite fruitful. It enabled Keynes to incorporate liquidity and the volatility of investment as essential parts of his analysis. It enabled Lachmann and Shackle to go beyond Hayek in the analysis of the dissemination of information in markets which Hayek started in his paper 'Economics and Knowledge' in 1937. Market participants do not make decisions on the basis of the information conveyed by prices, as Hayek implied, but on what they expect the future to be like on the basis of such information. Information conveyed by prices is first interpreted and reworked in millions of volatile, idiosyncratic minds and their divergent (i.e. differing) expectations in turn create new market information which is again reworked and so on. Extended in this way, the analysis allows new insights into the stability of markets and into the vast differences between, for example, asset markets and markets for consumer goods. These are the insights Lachmann tried to convey in his book on market processes of 1986.

> We are pleading for another thrust towards subjectivism by way of the introduction of the notion of the market as a process into the central area of economic theory. We believe that this notion would provide us with the means for a deeper understanding of the ways in which a market economy functions, for instance, by enabling us to appreciate the kaleidic nature of the network of plans that underlies all action.[121]

But there is a drawback in thus embracing the subjectivism of the Kaleidic society – one, incidentally, that Menger had already been well aware of back in 1871. If economic theory seeks its explanations purely in the actions of individuals and if economic theory is regarded as in some sense a reflection or representation of real economies which allows answers, in the manner of comparative statics, to questions of the type 'What would happen if this or that were done?', *then* any theory embracing the subjectivism of the Kaleidic society can, by the very nature of its premises, have only one answer, namely, 'Anything could happen'. This is, of course, not very helpful. After years of trying to give substance to the notion of a bounded rationality, Shackle appeared finally to have come to this conclusion as well when he was interviewed by Richard Ebeling for *The Austrian Economics Newsletter* in 1983.[122] He accorded the economist, at least qua economist, not much of a role at all. In response to a question, he said:

> I think there is pretty complete indeterminacy. I did spend a lot of energy trying to see if I could devise any theory of how expectations are formed, and I ended with the conclusion that expectations are far too elusive and subtle to find out any principles or rules to explain their emergence.

A little later in the interview he was asked: 'Given what you have said, what should economists do?' He said:

> I think they should give up giving advice, except on the most hesitant, the broadest grounds. I think they should introduce an ethical element, a more than ethical element. If a man is asked whether public expenditure should be cut or not, he perhaps should say, 'Well, if we cut it, we shall cause a great deal of misery; if we don't cut it, we don't know what the consequences will be, but we can't at least have this misery on our consciences.'

Lachmann did not follow his friend to these conclusions, though the questions perturbed him considerably. His colleagues and friends in New York did apparently (and presumably in good spirit) dub him a nihilist. But, while he liked the label *Radical Subjectivist*, he did not consider himself to be a nihilist. There was another dimension to his intellect which gave him a quite different perspective on the question of what economists should do.

Appendix 2: Value and demand: subjective and objective theories

Why are motor cars more expensive than bicycles? A common-sense answer to this question appears to be that more human effort is required to produce a motor car than a bicycle. But if the amount of effort spent in production is the sole criterion for how expensive (or valuable) things are, why does wine become more expensive (or valuable) simply by lying quietly in some cellar without any further human effort being spent on it, or why are some paintings extremely valuable even though the efforts of the artist may have been very meagrely rewarded when the paintings were first sold?

Questions concerning value have occupied many minds through the ages: What makes a thing valuable? What makes one thing more valuable than another? Is value an intrinsic property of valuable objects or is value, like beauty, in the eye of the beholder? Many theories of value have been put forward. They have contributed greatly towards the development of economics as a distinct discipline and especially towards the demand and supply analysis discussed in the previous chapter.

a. Value and price

One would be inclined to say that a very expensive piece of jewellery is valuable and, more generally, that any article which has a high price also has a high value. This suggests that whenever we speak of price we could as well speak of value. Under the conditions with which most of us are familiar, this suggestion is largely correct. For most purposes we may regard the price of a good as the value of that good, provided that the price in question is a market price. Not all prices of course are market prices.

Sometimes, however, we have to distinguish between value and price. Let us briefly consider one context in which this distinction is necessary. Many tribal societies have what are called subsistence economies. The term 'subsistence' is not meant to indicate that people in these societies always produce only for bare subsistence, i.e. barely enough to stay alive. Rather, it indicates that each household, consisting of the extended family of possibly many people related to each other, produces virtually everything it needs and produces only for its own needs. In other words, with the exception of a few items such as perhaps salt, the household does not buy anything, nor does it sell anything. Under such conditions there are of course hardly any markets and very few things have prices. But this does not mean that one cannot speak at all of value in a subsistence economy. Members of households may attach value to things because they find them useful or simply to their liking or because their production has involved the household in much toil and trouble. Such values may manifest themselves, for instance, in the trouble taken to protect possessions against the weather, fire or thieves. It is because certain things have value, even in a subsistence economy without prices, that they are called goods or possessions in contrast with mere things like unwanted stones and weeds in the veld.

The case of the subsistence economy and other cases one could mention suggest that a good has a value irrespective of whether or not it is traded and has a price. This in turn suggests that value is really something quite distinct from price. At one time, in fact, it was commonly thought that a good has an intrinsic value which is as much a property of the good as its shape, size, colour and so forth. If the good is traded, it was thought, the market price of the good need not always truly reflect the intrinsic value.

With certain exceptions (discussed in a much later chapter), economists nowadays consider the idea of intrinsic value to be quite mistaken. The value of a good is not an intrinsic property but an estimation in the minds of those who may want to use the good. How highly a person values his possessions and those he may acquire is entirely a private matter. In the absence of markets and prices, there would be no more to the question of value than this. But in markets a multitude of individual valuations come to be reflected in a market price that is the same for all. A theory of price must show how this happens and, as we shall see, demand and supply analysis does exactly that.

b. Usefulness and cost of production

Which affects value and price more: the usefulness of an article or its cost of production? It is often taken for granted that prices reflect costs of production. When prices rise, for instance, it is often presumed that costs of production had risen. In very many cases, costs of production is in fact a good indication price, and especially of a difference in prices. Motor cars are more expensive than bicycles and houses are more expensive than tents. But it is also true that on most occasions motor cars and houses are more useful than bicycles and tents. There are cases, such as those of old wine and works of art, in which usefulness or desiredness seems to exercise a more decisive influence over price than does cost of production. There are other cases. Towards the end of winter, for instance, ladies' winter-fashion garments are often sold in sales at prices which may be well below cost of production.

Both the usefulness of an article and its cost of production play a role in establishing its price. What has to be explained is just what these roles are. We shall see that demand and supply analysis also does that. The previous chapter has shown that market prices are established by the interplay of demand and supply. We now have to see that something like usefulness lies behind demand and cost of production (in relation to demand) behind supply. In order to understand demand and supply analysis more fully, we shall therefore, as it were, look behind the demand and supply curves. In this and the following chapter we shall concentrate our attention mainly on demand. Supply conditions will be dealt with thereafter.

The modern theory of price has grown out of the theories of value and price of the past. We shall appreciate the modern theory better if we briefly consider some of these past theories and the difficulties they encountered. To this we now turn.

c. The paradox of value

What has the usefulness of a good to do with how much one has to pay for the good? For a long time, it was thought that the answer to this question was: 'Very little'. So, for instance, it seemed to Adam Smith, one of the great names in the history of economics. In *An Inquiry into the Nature and Causes of the Wealth of Nations* (1776), he mused upon what has come to be called the paradox of value. Why is it, he wondered, that water is so very cheap, even though it is of such great use to mankind, while diamonds, which Adam Smith regarded as frivolous objects of little use, are so very expensive? In a similar vein, people pointed out how useful steel, coal and timber had been to mankind and how cheap they nevertheless were in comparison with gold and silver, which had much more limited uses.

On considerations such as these, a distinction was made between value in use and value in exchange or, more briefly, between *use value* and *exchange value*. To say that an object is valuable was held to be ambiguous. One could mean that the object was very useful or that it fetched a high price in the market. Use value and exchange value seemed to be almost completely unconnected. It was analysed, of course, that no one was likely to be alike to sell an article and hence to produce an article for sale, if that article had no use value whatsoever. But beyond that there seemed to be no connection at all.

The modern theory of price arose out of attempts to show that there was indeed a connection between use value and exchange value, i.e. that there was not really a paradox of value. The distinction between use value and exchange value is therefore no longer as important as it once was, though it is not without interest. For instance, Volume I of Karl Marx's *Das Kapital* (1B67) begins with a discussion of use value and exchange value and accepts the view that the two are unconnected. To this day, one of the broad Marxist critiques of capitalism is that capitalist production is motivated by profits and hence by exchange value, whereas it seems more desirable and sounder in the long run that the motivation for production should be the creation of use value, as it would be, so Marxists believe, under socialism.

d. The labour theory of value

When people believed that the usefulness of a good had little to do with the exchange value of the good, they naturally asked themselves what it was that determined exchange value. Not surprisingly perhaps, people usually tried to explain differences in exchange values by ascribing them to differences in costs of production. A special case of this type of explanation became very prominent and is generally known as the labour theory of value. The idea that there is a direct link between value and labour, with its connotation of strenuous effort and hard honest toil, has much appeal and, superficially at least, is quite plausible. But attempts to explain economic questions in terms of the labour theory of value were beset with many difficulties. There were also a number of versions of the theory.

The philosopher, John Locke, writing in the 1690s, argued that the laws relating to property should reorganize the rights of labour. When a person, as he put it, mixes his labour with what Nature has provided for all in common, then a property right is (or should be) created. Thus, the fish in the oceans do not belong to anyone, but when fishermen take the trouble to catch them, the fish become their property.

Locke's remarks on property do not constitute a theory of value. But the notion (which did not originate with Locke) that labour is the only source of value, and that therefore the entire product rightfully belongs to the workers, later became an important feature of the labour theory of value. We may refer to it as the *source of value* version of the theory. Among those who still adhere to a labour theory of value in our own day, this version seems to predominate, possibly because of its strong moral and political appeal. It must be pointed out, though, that Locke had a small-scale business in mind and included in 'labour' what we now separate out as the enterprise of the owner or capitalist, management, and manual work. When 'labour' is understood in the narrower sense (as is usual today) Locke's remarks are less convincing.

When Adam Smith wrote on the subject almost a century later, he took a different line. He did not regard labour as the source of value but rather as the ultimate and most meaningful measure of value. The real value of anything, he said, is the toil and trouble of acquiring it. Similarly, the real value of an asset is the toil and trouble it can save one if one exchanges it for necessary goods which one would otherwise have to acquire by one's own labour.

He also introduced a slightly different version of the theory in his well-known beaver and deer example. If it took twice as much labour to kill a beaver as to kill a deer, two deer would exchange for one beaver. In other words, the value of a good depends on the labour content of the good. He made it plain, however, that he thought this would apply only in a primitive state of society in which there was no machinery and equipment, and the land was not owned by anyone. Under such conditions, there would be no factors of production other than labour.

The labour theory of value is often associated with David Ricardo, who was second only to Adam Smith as a founder of economics. Ricardo was concerned to find an invariable measure of value and thought that labour time could serve as such. Writing in 1817, he refined what he took to be Adam Smith's value theory and thereby developed a rigorous *labour-content* version around which he built an elaborate theory of the whole economy. According to the labour-content version of the theory, exchange values are proportional to the quantity of labour, measured in man-hours, that went into the production of various goods. Ricardo saw that there would be problems if the theory were applied in cases where the ratio of labour to capital goods, the durability of capital goods and the duration of production processes varied much from industry to industry. He could not find a solution to these problems.

Later in the 19th century Karl Marx took over Ricardo's labour-content version of the theory and combined it with the source-of-value version. In Marx's theory, the intrinsic exchange value of a good is given by the labour time 'socially necessary' (i.e. required under the prevailing state of technology) for the production of the good, though prices are not necessarily proportional to exchange values. The cornerstone of Marx's analysis of the capitalist economy is that workers are exploited by capitalists. This idea of the exploitation of labour was derived from the source-of-value version of the theory. Since firms in a capitalist economy make profits and pay interest and rent, none of which usually ends up in the pockets of workers, there must be some others, namely capitalists, who are taking a cut from what rightfully belongs to the workers. This is the exploitation of labour.

It is in this form that the labour theory of value still lives on today among people with radical views and as something like the official doctrine of what we in the West call communist countries. In the Soviet Union it is illegal for a private citizen to operate, for instance, a taxi service with salaried taxi drivers. Such a private business would be held to involve exploitation of labour.

In other circles the labour theory of value was gradually abandoned and long ago lost its influence. There are immense difficulties in the application of the theory. There are the obvious difficulties created by the fact that occupations differ in the irksomeness of the work, that workers differ greatly in skill and expertise, that some workers are slower or lazier than others, and so on. But these are not the worst. There is also the problem which already worried Ricardo, namely, that prices seem also to be affected by the use of land and capital goods and by the enterprise of businessmen. Are these not productive factors on a par with labour? In Marxian theory this problem is ruled out because labour is proclaimed to be the sole source of exchange value and the problem becomes one of showing how prices are related to exchange values. But the notion of labour as the only source of value is more a matter of faith than anything else and is not always entirely convincing.

e. Subjective value theory

In the early 1870s there appeared three books which, over the following 25 years, were to inspire a new theory of value that changed the character of economics quite fundamentally. The authors of these books were Stanley Jevons in England (1871), Carl Menger in Austria (1871) and Leon Walras (1874), a Frenchman at a Swiss university. Working quite independently of each other, they showed how individual estimations or evaluations of the usefulness or desiredness of an article affected its market price. In effect, they showed that there was no paradox of value after all and that *use of values* and *exchange* values were not really unconnected. The three writers mentioned above were in fact not the first ever to have shown this and in the case of Menger and Walras it was not the main point of their books. But for some reason, this explanation of value gained wide acceptance after the publication of these books, albeit that it took a long time. Since individual estimations or evaluations may be regarded as being subjective, the new explanation of value was referred to as subjective value theory.

Beginner students of economics are usually introduced to a form of the theory called the marginal utility theory. Unfortunately, it is often hard to see the wood for the trees when one is dealing with the more formal aspects of this theory. It is quite easy to miss the point that the formulators of the theory wanted to make. Before we tackle the theory itself, therefore, we shall consider the insight that led to the theory and also to a solution of the paradox of value. It is really a very simple insight.

Let us recall that people had been impressed by the fact that certain very useful commodities such as water, steel, coal and timber were comparatively cheap while commodities with much more limited uses such as diamonds, gold and silver were very expensive. They concluded that value in use and value in exchange were simply different. In reaching this conclusion, they obviously had in mind the myriad uses which mankind in general had found for each commodity. Subjective value theory was to indicate that this was a misleading way of regarding usefulness when it was to be compared with price. The simple insight of the subjective value theorists was that the buyer and seller of a commodity were not at all concerned with how useful the commodity in general had been to mankind in general, but only with how useful they themselves in their particular circumstances found the particular quantity of the commodity, the purchase and sale of which they were contemplating.

The difference to which attention is being drawn here may be brought out more clearly by means of a rather extreme illustration. Imagine that you are walking down a busy city street when someone comes up to you and others, to sell you a bucketful of water. The chances are that you would decline the offer as would others in the street to whom the same offer is made. There would be no sale and hence no price would be established for that bucketful of water. But this does not mean that water in general is not useful to mankind in general, nor even that those who would decline the offer do not find water useful on other occasions. It means only that that particular bucketful of water would not be very useful or, put another way, that no one on that particular occasion would value a bucketful of water highly enough to be prepared to pay for it. Thus, it is the usefulness of the particular units of a commodity to be bought that is relevant to the price and not the usefulness of the commodity in general.

This distinction is important where a commodity has a great variety of uses, some vital and some not so vital. Water is used for drinking, cooking, having baths, washing clothes, and also for sprinkling lawns, filling swimming pools, washing cars, preventing dust on dirt roads, and so on. When water is plentiful and cheap, it may be used for all these purposes. If water becomes expensive, there will be some less important applications in which water is judged (or evaluated) to be not useful enough to make it worth paying the higher price (e.g. gardens are watered less often). If the price rises further, even more of the previous uses of water will be dispensed with, leaving only the more important uses. What this example shows is that the current price of a commodity should be related to its usefulness in its *least* important current application. The mistake made by those who posed the paradox of value (e.g. water is so cheap and yet so useful), is that they related price to the usefulness of a commodity in *all* its applications, including the most important ones.

The above type of reasoning has been formalized into the marginal utility theory. As it has developed, it now constitutes a theory of demand. The concept of demand is of course much older, but for a long time, economics lacked a satisfactory theory or explanation of demand. The marginal utility theory filled this gap. But of course, prices are established by both demand and supply, and therefore the marginal utility theory, as an explanation of demand, is only one part of a theory of price.

f. The marginal utility theory

Utility

In ordinary English, the word 'utility' means 'usefulness' and that is more or less the sense in which the word was first used in economics. However, since utility was used in the explanation of the demand for consumer goods (as distinct from capital goods and goods used as inputs in production) and since consumer goods are not so much useful as wanted or desired, 'utility' began to acquire a different meaning. It now means something like a measurable quantity of satisfaction.

One should not be put off by the idea of satisfaction as a measurable quantity. While there have been many not very successful attempts to measure utility in the specialized sense, it is not necessary for the purposes of the theory that it should actually be measured. Utility is merely an explanatory device used to show the relation between individual evaluations and market prices. In the next chapter we shall see that we can explain this relation in a way that does not involve measurable opportunities of satisfaction at all. We shall merely have to assume that consumers have preferences. For the moment, however, we shall speak of the utility of milk, shoes, matches and so on, meaning by that the satisfaction or utility these things provide to a consumer.

Marginal utility

We saw a page or two ago that one can put the various uses of a commodity into an order of importance. The point was illustrated by the example of water, but many other examples are at hand. Let us consider the petrol used in a private car. Some trips made in the car may be regarded as essential, others merely as quite important and some excursions may be judged to be quite trivial. When the price of petrol rises, the use of the car may be curtailed in so far as some of the less important trips are dispensed with. Should the price of petrol fall, the use of the car may be extended. (We are here ignoring the costs involved in the wear and tear to tyres and other parts of the car.)

The *least* important of the *current* uses of a commodity such as petrol is called its *marginal* use, i.e. it is at the margin of all its uses. The quantity of a commodity involved in the marginal use is called the marginal unit of the commodity. We often say that the marginal unit of a commodity is the last one used or, in some contexts, that it is the extra, additional or incremental unit of a commodity. In this way we can also speak of the marginal unit of a commodity which appears to have only one use, as for example toothpaste. The second, third and further additional tube of toothpaste has a very limited current use. In this particular case, we could say that its current use is to be stored away for brushing teeth at a later time. In an analogous way, additional loaves of bread, bars of chocolate, bottles of soft drink, and so on have even less urgent uses. In each case, the marginal unit of the commodity is that unit for which there is the least important use.

The utility which a consumer or his household derives from or ascribes to the marginal unit of a commodity is called the *marginal utility* of the commodity. Thus, we speak of the marginal utility of water, petrol, toothpaste, bread and so on. It should be clear by now that marginal utility is not an inherent quality of a good, but rather the use value which a particular consumer or his household ascribes to the marginal unit of a commodity, i.e. to a commodity in the use judged to be the least important. Marginal utility is an extremely important concept. It allows us to express, in our theory, individual evaluations by consumers regarding the usefulness, or better, the wantedness or desiredness of the goods they use.

The reason for expressing the theory in terms of marginal or incremental units is that it lends itself in the form to treatment in terms of the calculus. Readers who are familiar with this part of mathematics will be able to see this easily. The calculus provides us with techniques for stating, for example, the conditions for maximizing certain quantities under certain constrains and thus allows a considerable refinement of the theory.

Diminishing marginal utility

Let us imagine a consumer who acquires successively more units of a commodity, that he acquires them one by one, so that each unit is at some stage the marginal unit, and finally that at each stage he makes a note of the utility he ascribes to the marginal unit, i.e. of the marginal utility of the commodity. With the information collected by one imaginary consumer, we are in a position to draw up a table showing the marginal utilities and the total utilities the consumer ascribes to the commodity in question (see Table 1). The figures are purely hypothetical and could relate to any consumer good – we shall refer to it merely as commodity x.

Table 1: Schedule of marginal and total utilities

Number of units of commodity x at the disposal of the consumer	Marginal utility of commodity x	Total utility of commodity x
1	40	40
2	35	75
3	30	105
4	20	125
5	5	130
6	3	133
7	2	135

The figures in the marginal utility column are those which are imagined were jotted down by the consumer. The figures in the total utility column are the cumulative totals of the marginal utilities. Thus, when the consumer had four units of commodity x, the total utility of commodity x was 40 + 35 + 30 + 20 = 125. (We could also have imagined that the consumer jotted down the figures for total utility. We could then have derived the marginal utility column by taking the differences between successive figures for total utility. It does not really matter which approach is adopted.) The total utility of a commodity is therefore the utility a consumer ascribes to all the units of the commodity at his disposal whereas the marginal utility of the commodity is the utility ascribed to the marginal unit only.

If we look down the marginal utility column, we see that the numbers become progressively smaller. This is the point which the table is meant to illustrate, and any other set of figures with this feature would have done as well. As explained already, utility is merely an explanatory device, and it is not suggested that any consumer does actually draw up such a table. Our table is meant to illustrate the central assumption of marginal utility theory, namely, what is usually referred to as the law of *diminishing marginal utility*. This law states that the *marginal utility* of a commodity diminishes as a consumer acquires additional units of the commodity. No assumption is made about how rapidly marginal utility diminishes or how low it may fall – it could fall to zero or even become negative. The law states only that marginal utility diminishes with increasing quantity.

If a consumer could provide us with a sequence of figures for either marginal or total utility (as in Table 1) for every consumer good he may consider buying, we would have a representation of the consumer's *tastes*. This representation of what even in ordinary language is called a person's tastes is not as satisfactory as one given in the more advanced treatment of the subject in terms of revealed preferences, but for our present purposes it will do. The tastes of different consumers may of course differ markedly, but they are taken to have at least one common feature, namely, that they conform to the law of diminishing marginal utility.

How tenable is the law of diminishing marginal utility? Actually, a little thought shows that it merely formalizes something with which we are quite familiar. We know that while there may be no end to the variety of uses we have for some goods, there is a limit to the enjoyment we derive from any one use of a commodity. If bread were given away free, we would not gorge ourselves until we burst. Of course, a person with a surplus of bread to eat may find other uses for it, such as giving it away to the needy. In so far as he derives satisfaction from making others less hungry, bread retains a positive marginal utility for him. Similarly, in our previous examples of water and petrol, we took it for granted that a point of satiety is soon reached in any one use of these commodities, but that there are then further uses. It seems reasonable to suppose that people put the various uses of a commodity into an order of importance and further that a person with very little of a commodity meets his most urgent needs with it (i.e. the most important uses). As he acquires greater amounts of the commodity, he proceeds towards ever less important uses. This is really all that we are expressing when we say that the marginal utility of a commodity diminishes as a consumer acquires additional units of the commodity.

With the terminology now at our disposal, we are in a position to state more easily why there is not really a paradox of value. The paradox was: water is very useful, yet it is cheap; diamonds are trivial items, yet they are expensive. We may paraphrase the observation that water is very useful as follows: For the vast majority of mankind, water has a high total utility. We may agree with this statement but have to point out that it is not total, but marginal utility that should be compared with price. In most reasonably affluent societies, people have a plentiful supply of water so that the marginal utility of water is low. Most people do not have an urgent use for an extra litre of water and would not be prepared to pay much for it. Since all units of water are normally sold at the same price, water must be cheap.

One can easily imagine a situation in which the marginal utility of water would be extremely high, and a small quantity of water would be valued more highly than a diamond. Let us imagine the following: A man finds a diamond as big as an acorn on the Skeleton Coast. A while later, he discovers that his water bottle has sprung a leak and he is without drinking water. Being in a desolate place between the Namib desert and the sea, he is destitute. When he is on the verge of dying of thirst, he would be only too glad to exchange the diamond for a glass of water.

Consumer equilibrium

A consumer has a wide choice in spending his income. He has to decide which goods to buy and how much of each. One may presume that he tries to allocate his expenditure to various goods in such a way that he makes the most of his money, as the common expression goes. We now want to use the marginal utility theory to capture more formally this everyday idea of making the most of one's money or of allocating one's expenditure in the best possible way.

Let us reflect for a moment on what is implied by these colloquial expressions. First, it is implied that one has limited financial resources, for otherwise there would be no need to economize or to give any thought whatever to how well one organizes one's expenditure. A consumer normally has a certain income. He may have some savings and perhaps even some inherited assets. He may also be able to borrow money, but there is a limit to how much he can spend. For short, we shall say simply that a consumer has a limited income. Secondly, making the most of one's money implies that one cannot improve on the way one's limited income has been allocated. One cannot spend a little less on one type of good and a little more on another and regard the result as an improvement. A consumer makes the most of his limited income when he cannot improve on the way that he allocates his limited income to various goods.

Let us rephrase this in terms of the marginal utility theory. We shall say that a consumer is in equilibrium when he cannot gain further utility by altering the way he allocates his limited income to various goods. It follows that a consumer is in equilibrium when he maximizes overall utility, where overall utility is the sum of the total utilities he ascribes to all the goods at his disposal.

We now want to find a way of stating the conditions under which overall utility is maximized. This is really quite easy. In equilibrium the consumer cannot gain utility by reallocating, say, 1 Rand of expenditure from one type of good to another. In other words, the utility to be obtained per Rand is the same in the case of every type of good. We must be careful here to allow for the fact that goods are not equally expensive. The utility to be obtained per Rand (or other monetary unit) of expenditure can be found by dividing the utility of a marginal unit, i.e. the marginal utility, by the amount that has to be paid for the marginal unit, i.e. by the price of the commodity in question. Overall utility is therefore maximized, and the consumer is in equilibrium when the ratio of marginal utility to price is the same in the case of every commodity.

We may therefore state the conditions for consumer equilibrium in the following way: A consumer is in equilibrium when he allocates his limited income to various goods in such a way that:

$$\frac{MU1}{P1} = \frac{MU2}{P2} = \frac{MUu}{Pu} \qquad (1)$$

Where *MU*1, *MU*2, …, *MU*u are the marginal utilities of the u commodities bought by the consumer and *P*l, *P*2, …, *P*u are the prices of the u commodities bought by the consumer.

How does a consumer reach equilibrium? Suppose that on his initial plan of expenditure a consumer is not in equilibrium. There must then be at least two commodities x and y, among the u that he plans to buy, such that (*MU*x/*P*x) (*MU*y/*P*y). If the consumer were now to reallocate his proposed expenditure in such a way that he would spend less on commodity x and more on commodity y, he would give up less utility than he would gain (because utility per Rand on x is less than utility per Rand on y). As long as the inequality holds, the consumer can increase his overall utility by substituting y for x. However, according to the law of diminishing marginal utility, as the consumer acquires more of y and has less of x the marginal utility of x rises. A point must therefore be reached where (*MU*x/*P*x) = (*MU*y/*P*y). When this point has been reached in the case of every pair of commodities from among the u commodities in the consumer's budget, there is no further scope for increasing utility by a reallocation of expenditure. Overall utility is at a maximum and the consumer is in equilibrium.

Consumer equilibrium is a theoretical representation of the weighing up at the margin which is quite consciously carried out when provision is made for the future. One may imagine the equipping of an Antarctic expedition. After the most obvious items have been assembled and funds are short, there will be a careful weighing up whether it is more worthwhile to take additional jerseys or additional torch batteries or additional spare parts for the stoves and so on. Few people may budget so carefully. But it is assumed that the individual evaluations and considerations made, say, by a housewife in a supermarket are similar in principle.

The determinants of the quantity demanded

Let us consider *any* two commodities x and y among those bought by a consumer.

It follows from equation (1) above that in consumer equilibrium:

$$\frac{MUx}{Px} = \frac{MUy}{Py}$$

Where *MU*x and *MU*y are the marginal utilities and *P*x and *P*y the prices of commodities x and y respectively. We can rearrange this equation and write it as follows:

$$\frac{MUx}{MUy} = \frac{Px}{Py} \qquad (2)$$

Formulated as in equation (2), the conditions for consumer equilibrium can give us an idea of the factors that determine how much of a commodity a consumer demands. Suppose that a consumer is not in equilibrium. Equation (2) would then not hold. The ratio of marginal utilities would be either greater or smaller than the ratio of prices. What can the consumer do to reach equilibrium? The ratio of prices or, as economists say, the relative prices shown on the right-hand side of equation (2) are usually beyond the control of the individual consumer. Prices do change of course, but there is usually nothing that the individual consumer can do to change prices. To attain equilibrium, he must also do something which has the effect of changing the ratio of marginal utilities on the left-hand side of equation (2).

We have already seen that a consumer attains equilibrium by reallocating his. expenditure. In accordance with the law of diminishing marginal utility, he can decrease the marginal utility of a commodity by buying more of the commodity or increase it by buying less. He has to do this under the constraint of a limited income, so that buying more of some commodities involves buying less of others. He therefore attains equilibrium by adjusting his expenditure on various goods until, in the terms of equation (2), MUx/MUy becomes equal to Px/Py, i.e. to the relative prices over which he has no control.

Such adjustments of expenditure amount to changes in the quantities of various commodities that the consumer demands. In order to find the determinants of these quantities demanded, we must look again at equation (2) to see what kind of changes disturb consumer equilibrium so that expenditure has to be adjusted to fit in with relative prices. Obviously, a change in relative prices is one such change. Looking at the other side of the equation, we see that a change in the consumer's evaluations would disturb the equilibrium because marginal utilities would change independently of any quantity adjustments. We call this a change in the consumer's tastes. Finally, the amount that can be bought quite obviously also depends on the amount that can be spent.

We may therefore list the following determinants of the quantities of various commodities demanded by consumers or, for short, the following determinants of demand:

(a) consumers' incomes
(b) consumers' tastes
(c) relative prices.

Price–quantity relations

Relative prices as determinants of demand need further explanation. That the quantity of beef demanded varies with the price of beef is something with which we are all familiar. But the quantity of beef demanded may also vary because beef has become relatively cheaper or relatively more expensive, even though the price of beef expressed in Rand and cents has not changed at all. There are familiar cases involving substitutes. The quantity of beef demanded may rise because the price of mutton has gone up so that beef has become relatively cheaper than mutton. The marginal utility theory tells us that quantities demanded may be affected by any change in relative prices, not only by one that involves substitutes or complementary goods, although in many cases the effect is probably very small.

Let us write down equation (1) again.

$$\frac{MU1}{P1} = \frac{MU2}{P2} = \frac{MUu}{Pu} \tag{1}$$

We shall consider the i-th commodity, which is any one of the u commodities. Suppose that Pi rises so that the numerical value of the ratio (*MUi*/*Pi*) is now less than the numerical value of the ratios for the (u-1) other commodities. Since the consumer cannot change *Pi*, he has to regain equilibrium by increasing *MUi* and he can do this by buying less of the (i-th) commodity. If he spends the same amount on the (i-th) commodity as he did before its price rose, he would of course be buying less of it because it is now more expensive. If this reduction in quantity raised *MUi* just sufficiently to restore equilibrium, the consumer's demand for all commodities other than i would be unaffected by the rise in *Pi*.

This case is possible but not very likely. It is the case where the price elasticity of the consumer's demand for the i-th commodity between the old and the new price is exactly unity. It is one possibility out of an infinite number of possibilities. The chances are that the elasticity of the consumer's demand is either greater or less than 1. Whether it is elastic or inelastic (greater or less than 1) depends upon how rapidly marginal utilities diminish with increase in quantity and rise with decrease in quantity. We have made no assumption about this feature of consumers' tastes. It could vary from consumer to consumer. We have assumed only that consumers' tastes conform to the law of diminishing marginal utility, i.e. that marginal utilities do diminish with increase in quantity, and nothing about the rate of diminishing.

If his demand for the i-th commodity is inelastic (high rate of change of marginal utilities), the consumer would spend more on it in the new equilibrium than he did in the old equilibrium before the price rose. Because his income is limited, he would have to spend less on other commodities. If his demand is elastic (low rate of change of original utilities), he would spend less on the i-th commodity in the new equilibrium than he did in the old equilibrium. Since he always spends his entire income – even if it is partly on future goods in the form of savings – he would spend more on commodities other than the i-th. When the consumer accordingly spends either more or less on a commodity of which the price has not changed, the numerical value of the ratio (*MU/P*) for that commodity changes. Since in equilibrium, as we can see from equation (1), the numerical value of this ratio is the same for every commodity, a change in *P*i induces the consumer to demand changed quantities of not only some but all commodities, except in the unlikely case of his demand for the i-th commodity being of unitary price elasticity. We are therefore led to believe on purely theoretical grounds *that a change in the price of any one commodity affects a consumer's demand for all commodities*. There is, however, one price-quantity relation which is of special interest, namely, the relation between the price and quantity demanded of the same commodity. This is of course the price quantity relation singled out and represented by the demand curve. It is of special interest because the effect of price on quantity is in this case likely to be strong and because we can be more definite about the nature of this kind of price–quality relation than about that of others. We saw above that we do not know whether a change in price induces a consumer to demand greater or smaller quantities of commodities other than the one of which the price has changed, only that there is very likely to be some effect. In the case of the price and quantity of the same commodity, the effect is more predictable.

If we look carefully at equation (1), we see that a rise in the price of a commodity always leads a consumer to demand less of that commodity and a fall leads him to demand more. Take *P*i, *MU*i and Qi as the price, marginal utility and quantity demanded of the i-th commodity respectively. If *P*i rises, the value of (*MU*i/*P*i) falls. If the consumer were *not* to reduce Qi to raise *MU*i, he would have to restore equilibrium by buying more of all other commodities to reduce their marginal utilities. But he does not have enough income to do this. If *P*i falls, the value of (*MU*i/*P*i) rises. If the consumer were *not* to raise Qi to reduce *MU*i, he would have to buy less of all other commodities to raise their marginal utilities and to bring about equality of all the (*MU/P*) ratios. But he would then not spend all his income. He would not maximize his overall utility and would not be in equilibrium.

An inverse relation between the price of a commodity and the quantity of the commodity demanded by a consumer can therefore be deduced from the law of diminishing marginal utility and the conditions for consumer equilibrium. The negative slope of the usual market demand curve – its slope downwards from left to right – represents the same inverse relation between price and quantity, but in the market as a whole. Previously, we called this inverse relation in the market the law of demand. Since market demand is the combined demand of all market participants, we can now see *that marginal utility theory gives us confirmation of the law of demand on purely theoretical grounds.*

The individual consumer and the market

Marginal utility theory deals with the individual consumer. It deals with one consumer's demand for many commodities whereas demand and supply analysis involves the demand of many consumers for one commodity. However, it is not difficult to trace the connection between individual evaluations as they appear in marginal utility theory and market demand as it appears in demand and supply analysis. We may consider one consumer's demand for one commodity by drawing an individual demand curve representing the relevant price quantity relations. In doing so, we have to remember that the individual demand curve for the i-th commodity has to be redrawn whenever there is a change in the price of a commodity. Other than the i-th, because, as we have seen, such change is likely to affect the demand for the i-th commodity.

If we have the individual demand curves of all the buyers in a market, we can derive the market demand curve by horizontal summation. This means that we take the sum of the individual quantities demanded at a price and plot it against that price on a diagram showing the market demand. We repeat the procedure for every other price. Thus, when there are only three buyers in a market, the derivation of one point on the textbook market demand curve can be shown diagrammatically

There is really no difficulty in horizontal summation. Nevertheless, as we broaden our perspective from the individual to the market, some serious difficulties may arise. A perceptive reader may already have seen a difficulty when relative prices were described as determinants of the quantities demanded by a consumer. How does this fit in with the determination of prices described in the previous chapter? Do individual consumers adjust the quantities they demand to pre-existing market prices, or do they help to determine market prices by demanding certain quantities? Just what determines and what is determined?

The word 'determinant' in economics may mislead because it is used in a special way. Economists see the economy as a vast system of interrelations in which virtually everything depends on everything else. It is hard to grasp such a system of interrelations all at once. So, economists try to build up their understanding bit by bit by asking questions such as 'What determines market prices?' or 'What determines how much of various commodities or consumer demands?' They summarize their answers in equation form and call the independent variables in the equations the determinants of the dependent variables. They choose dependent variables according to the questions they ask, and the independent variables come in with the answers. One may therefore find, for instance, that prices sometimes appear as dependent variables and sometimes as independent variables, i.e. sometimes as being determined and sometimes as determinants.

The nature of what actually happens in the economy does not of course continually change to and fro according to the questions asked in economic theory. The questions are asked in order to develop theories which are to be building blocks for a model by which the actual economy is to be understood. The marginal utility theory and demand and supply analysis are merely two such building blocks. We shall ask further questions, e.g. about production and the prices of factors of production, and these will yield further theories and therefore further pieces of the model. Eventually, of course, all the pieces have to be filled together. We have to await the chapter on general equilibrium for an indication of how this may be done.

g. The wider implications of subjective value theory

There was more to the advent of subjective value theory than the marginal utility theory. The developments in economic theory which started in the 1870s are sometimes referred to as the Subjective Revolution in economics but more commonly as the Marginal Revolution. The latter term indicates just how important marginal analysis has proved to be. It has come to be applied in other spheres of economics so that we now speak of marginal costs, marginal products and so on. Marginal analysis, as pointed out earlier, opened the way to an application of the calculus in economics and more particularly to mathematical maximization and minimization techniques. This created a whole new style of economics which we now call micro-economics.

As a Subjective Revolution, the new developments were equally far-reaching. Economics acquired a theory of demand which at best the dominant strand of economics in the English-speaking world did not have before that. With the theory of demand came a new way of looking at economic problems. Scarcity and choice came to occupy the centre of the stage and economics to a large extent became – in the words of Lionel Robbins' well-known definition of the subject – the science which studies human behaviour as a relationship between ends and scarce means which have alternative uses. This definition first appeared in a book published in 1932 in which Lionel Robbins presented to the English-speaking world the views of the followers of Carl Menger, the Austrian who, it will be remembered, was one of three founders of subjective value theory.

Notes

1 L.M. Lachmann: '*An Austrian Stocktaking*' in Spadaro (1978, p 3).

2 Even Ricardo gave them their due mention. Ricardo (1817, second paragraph of the first chapter): 'If a commodity were in no way useful in other words, if it could in no way contribute to our gratification – it would be destitute of exchangeable value.' He went on: 'Possessing utility, commodities derive their exchangeable value from …'.

3 '*Theorie des subjektiven Wertes*' instead of '*subjektive Werttheorie*'. Oskar Morgenstern in Mises and Spiethoff (1931, p 10).

4 *Notice of a General Mathematical Theory of Political Economy*. Reprinted in Jevons (1957 [1871]) as Appendix III. The title was changed in 1866, the words '*Brief Account*' being substituted for '*Notice*'.

5 Jevons (1957 [1871]). The quotations are taken from pp xxvi, 21 and 23f. See also the remarks by R.D. Collison Black (1973) in Black et al (p 107f).

6 Jevons (1879a, p 532).

7 Jaffé (1976, p 519).

8 Jevons (1957 [1871], quotations from pp vii and 104.

9 Jaffé (1976, p 518). As will be pointed out below, Walras' theory nevertheless had a utilitarian aspect. Schumpeter (1952, p 119) credited Walras with a utilitarian outlook.

10 See Hutchison (1953, pp 197–199).

11 William Jaffé (1973, pp 122–126).

12 Walras (1873) '*Principe d'une theorie mathematique de l'echange.*' *Ibid.* p 114f.

13 *Ibid.* pp 126–132 and Jaffé (1976, pp 513–515). Jaffé, W. (1973) 'Leon Walras's Role in the "Marginal Revolution" of the 1870s', in Black et al (1973, pp 113–139).

14 Jaffé (1976, p 515).

15 Schumpeter commented on the irony of Pareto's position as 'patron saint' of modern welfare economics. 'The story of how, once more, he came to render a service to a cause with which he was – or would be – completely out of sympathy is not without its humour.' Schumpeter (1952, p 130).

16 Pareto (1953 [1900]). Certain extracts from these two letters are also discussed by V.J. Tarascio in Black et al (1973, pp 140–159).

17 Pareto (1953 [1900]). Quotations, in order, are from pp 204, 207 and 185.

18 See Walras (1874, p 71). Walras said of the sciences of which he took economics to be an example: 'From real-type concepts, these sciences abstract ideal-type concepts which they define and then on the basis of these definitions they construct a priori the whole framework of their theorems and proofs. After that they *go* back to experience not to confirm but to apply their conclusions.' To bear out his statement he said that the theorems of geometry do not need empirical confirm and would in any case be confirmed only approximately by experience. Walras also said (1874, p 61): 'A truth long ago demonstrated by the Platonic philosophy is that science does not study corporal entities but universals of which these entities are manifestations. Corporal entities come and go, but universals remain for ever. Universals, their relations, and their laws, are the object of all scientific study.' One may compare to this Pareto's remark: 'We experimentalists … leave the concrete, which we always looking at lovingly, only with reluctance and when forced by necessity' (Pareto (1953 [1900]), p 206). As we shall see presently, Walras' attitude to these matters was much closer to von Mises' and Menger's than to Pareto's.

19 Jevons (1874). The whole book is devoted to the purpose but see especially Chapters 23 and 24, pp 504–550.

20 See Pareto (1953 [1900], pp 201–207).

21 Jevons (1874, p 507) and Pareto (1953 [1900], p 205).

22 Jevons (1874, p 737).

23 Pareto (1953 [1900], p 207).

24 Menger (1871, pp 78–81 and 108–113).

25 See Schumpeter (1954, pp 300–302, 1054–1059 and Part II, *passim*) and also Kauder (1965, pp 15–29).

26 Mill (1863, p 1).

27 Schumpeter (1954, p 302; see also p 408).

[28] Referring to critics of utilitarianism, Mill said: 'To suppose that life has … no higher end than pleasure … they designate as utterly mean and grovelling; as a doctrine worthy only of swine, to whom the followers of Epicurus were, at a very early period, contemptuously likened; and modern holders of the doctrine are occasionally made the subject of equally polite comparisons. … When thus attacked, the Epicureans have always answered, that it is not they, but their accusers, who represent human nature in a degrading light; since the accusation supposes human beings to be capable of no pleasures except those of which swine are capable' (Mill 1863, p 7).

[29] Mill (1863, p 6).

[30] Bentham (1843, p 1f).

[31] Bentham (1843, Text, pp 17 and 202f).

[32] Bentham (1843, Intro, p 3).

[33] Bentham (1843, Intro, pp 22 and 23).

[34] Mill (1872 [1843], Book VI, Chapter VIII, para 3). Mill went on to say that in their '*political reasonings*' the word interest had to be understood to mean 'what is commonly termed private, or worldly, interest'.

[35] Jevons (1957 [1871], p 39).

[36] Mises (1949, p 120).

[37] Jevons (1957 [1871], p 45).

[38] Marshall (1890, p 93).

[39] Irving Fisher: 'Is "Utility" the most suitable term for the concept it is used to denote?' (1918).

[40] Schumpeter (1954, p 130f).

[41] Schumpeter (1954, pp 132–134, 408f, 830f and 1056f). Schumpeter called utilitarianism 'the shallowest of all conceivable philosophies of life'.

[42] Mises (1949, p 15; see also p 21).

[43] Mises (1933, p 142).

[44] Mises (1949, p 15).

[45] Mises (1949, pp 32 and 38).

[46] Cairnes (1874, p 15f).

[47] To a modern reader with the benefit of hindsight, Cairnes' way of putting this must seem quite neat. Unfortunately, the neatness is somewhat coincidental. It is evident from the seven pages in which Cairnes discussed the matter that he did not understand Jevons' theory. He had no inkling of what in a later textbook terminology is the distinction between total and marginal utility to a consumer. Cairnes said that he had taken every pain to understand Mr Jevons' doctrine (1874, p 13), which goes to show just how difficult it is to convey unfamiliar ideas even to sophisticated intellects.

[48] Schumpeter (1952, p 124).

[49] Pareto (1953 [1900]), p 190f.

[50] Mill (1872 [1843], Book VI, Chapter III; see also Chapters IV and IX).

[51] Hutchison (1953, p 16).

[52] Pareto (1953 [1900], p 207).

[53] Schumpeter (1952, p 126). Jaffé (1976) says of this passage that it does not accurately describe Walras' intellectual progress. It will become apparent below (if indeed an indication is necessary) that it applies even less to Menger.

[54] Hutchison (1953, p 16).

[55] Streissler (1973, pp 160–175).

[56] Streissler (1973). The quotations, in order, are from pp 160f, 166, 169 and 172.

[57] Streissler (1973, p 175).

[58] Menger (1871, p xlvi f).

59 Hayek in the introduction to the LSE reprint of the *Grundsatze* (Menger 1871, p ix). Hutchison (1953, p 148) where the extract quoted is from (Jaffé 1935, p 200). See also Kauder (1965, p 90f); Jaffé (1976, p 521); Lachmann (1977, p 48f) and D. Winch (1972, p 331) and E. Streissler (1973, p 174).

60 See, for instance, Menger (1883, p 76f).

61 T.W. Hutchison, 'Some Themes from Investigations into Method', in Hicks and Weber (1973, p 31). He was citing Emil Kauder who drew the conclusion from Menger's papers in Tokyo.

62 See Kauder (1965, pp 95–100) and T.W. Hutchison in Hicks and Weber (1973, p 118).

63 Kauder (1965, p 95).

64 Menger (1871, p xlv), repeated in Menger (1883, p 41).

65 Translation by Benjamin Jowett. No reference.

66 Menger (1883, p 14).

67 Menger (1883, pp 4, 12, 25–27, 41, etc.).

68 Menger (1889, pp 197–199).

69 Menger (1883, pp 5 and 33).

70 Menger (1883, p 4).

71 Menger (1871, p xlv).

72 Menger (1889, pp 189–192).

73 Menger (1883, p 260).

74 Menger (1871, p xlv).

75 Menger (1871, p xlvi).

76 Menger (1871).

77 Menger (1883, p 259).

78 Menger (1871, p xlvii).

79 Menger (1883, p xx).

80 Menger (1883, p 18).

81 Menger (1883: Chapters 4 and 5). In the preface to Menger (1871) he had said that the reduction to types etc. is properly called the empirical method.

82 '*Regelmassigkeiten in der Coexistenz und Aufeinanderfolge.*' Menger (1883, first mentioned on p 4 and then throughout the book). Sometimes (e.g. p 34) it was a regularity of phenomenal forms, which seems more correct, but usually he spoke of a regularity of phenomena, eventually of generally demarcated phenomena.

83 '*Volle empirische Wirklichkeit.*' Menger (1883, pp vii and 34f and thereafter). He also spoke of concrete phenomena.

84 Menger (1883, p 35).

85 Menger (1883, pp 34–38 and 46–48).

86 At the end of the *Posterior Analytics* (Book II, Chapter 19), Aristotle discusses induction, intuition and primary premises in an uncharacteristically vague manner.

87 Menger (1883, p 40). The principle states that whatever has been observed once would occur again under exactly the same conditions.

88 Menger (1883, p 6fn.).

89 Menger (1883, p 53f).

90 Menger (1883, p 42; see also pp 25 and 40f).

91 Menger (1883, pp 59 and 265).

92 Menger (1889, p 105). See also pp 190, 192 and 199, where he speaks of inner connection and inner causation (*innerer Zusammenhang, innere Verursachung*).

93 See Boehner (1957, pp xxviii f and 41–45); also Leff (1975, pp 78–123).

94 *Posterior Analytics II.* 2., translation by G. Mure. In one place, Menger (1883, p 87) states the maxim: *Scire est per causas scire.* See also note 94 above. No book title.

95 Menger (1883, p 59).
96 Menger (1883, p 54).
97 Walras (1874, p 71). After explaining how he understood the method of geometry, Walras continued: 'Following this same procedure, the pure theory of economics ought to take over from experience certain type concepts, like those of exchange, supply, demand, market. ... From these real-type concepts the pure science of economics should then abstract and define idea-type concepts in terms of which it carries on its reasoning. The return to reality should not take place until the science is completed and then only with a view to practical applications' (see also note 20 above). Menger may have had misgivings about the fact that phenomenal forms were described as concepts, but he could not have objected much to the way Walras put this. That their legacies are so different is probably due to the circumstance that Walras' theory was taken over by a man with strong and different philosophical convictions whereas Menger's value theory was established by men who appear not to have had strong philosophical views.
98 Menger (1883, pp 72–74).
99 Menger (1883/1960, p 84).
100 Hutchison (1973, p 22f).
101 Menger (1883, pp 75–77).
102 Menger (1883, pp 46, 56, 259f and 264).
103 See Borch (1973, p 61).
104 This is the third of four senses of the word empirical given in the *Oxford English Dictionary*. The first two are similar but applied only to medicine. The fourth is the sense common in economic discussion.
105 See Menger (1883, pp 78–80).
106 L.M. Lachmann: 'An Austrian Stocktaking' in Spadaro (1978, p 2).
107 Kirzner (1979, p 151).
108 Menger (1883, p 67).
109 Menger (1883, p 66f). It was said in the course of countering the view that economic questions should be dealt with in the context of the whole social and political development of a country.
110 Menger (1883, p 36f and Appendix V).
111 These are the words of a heading Hayek gave to a chapter explaining subjectivism in the sense under discussion here. See Hayek (1952, pp 25–35).
112 T.W. Hutchison in Hicks and Weber (1973, p 16).
113 Lachmann (1943a).
114 Lachmann (1950, p 217).
115 Lachmann (1976).
116 Lachmann (1982). See especially pp 36–39.
117 Mises, L. (1962, p 18).
118 Lachmann (1977) *Capital, Expectations, and the Market Process*, p 85.
119 The quotation to which Shackle refers in the letter appears in the second edition of his book *Decision, Order and Time in Human Affairs* (Cambridge University Press, 1969), p 41. The words quoted in the paragraph above are part of this quotation.
120 But not everything Keynes said and stood for. For instance he described Keynes' remarks on the rate of interest in the final chapter of the *General Theory* as imbecilic. Nor was he in sympathy with Keynes's welfare-state leanings.
121 Lachmann (1986, p 147).
122 The quotations that follow are taken from Shackle (1983, pp 6 and 7).

5

Anthropinism

1. Introduction

In this chapter we shall try to arrive at a conception of perceiving and knowing which will allow us in later chapters to form a clearer idea of how a knowledge of institutionalized conduct fits into our understanding of economic questions, as well as to investigate the allegation made in the Preface that the presuppositions on which micro-economic equilibrium theory is based lead to a neglect of such knowledge. We shall not try, of course, to resolve the age-old problems of the theory of knowledge, but rather to outline a synthesis of the subject matter of the previous chapter, though here and there novel features may perhaps be introduced.

We have seen that two major questions with which the theory of knowledge has had to contend were first how to reconcile being with becoming or things with events and secondly how to reconcile the general with the particular, or theories with facts. Some reflection on what was said in Chapter 1 will show that economics also has to contend with these questions within its own subject matter. The conception that is desired here must therefore allow us to handle these reconciliations and it must allow us to do this in the context of the subjective re-orientation of thought. What this amounts to is that we want to be able to conceive how particular events or changes are understood in terms of general and unchanging meaning. Here it must be borne in mind that the particular purchase or sale, the particular act of consumption or production, however it is treated statistically, is nevertheless an historical event.

Some specialized terminology cannot be avoided, but the guiding principle will be that unobtrusiveness is a virtue in such matters. Also since a conceptual framework is a whole made up of interconnected parts whereas an exposition necessarily proceeds from a beginning to an end, it is unavoidable that the explanation of a term should sometimes contain as yet unexplained terms. However, these will be sufficiently self-explanatory to make this a feasible procedure. The term *anthropinism* will be applied to the conceptual framework as a whole. Subjectivism would perhaps have been the most correct term, but it has many connotations which are not intended here. To some it seems to signify the rejection of even the possibility of knowledge and that is certainly not the intention here.

As an antonym of objective, the word subjective is often taken to indicate something idiosyncratic, emotive, arbitrary, or even imaginary, illusory or fanciful. None of these meanings nor the juxtaposition of objective and subjective is intended here. Within economics the word *subjective* is associated with certain theories of value and demand with which the present proposals have no affinity. The term subjectivism is also used by some economists to describe a point of view and an approach to economics which differ in certain important respects from what is proposed here, though there is also much common ground. To avoid misunderstanding, therefore, the obscure words anthropinism and anthropinistic have been pressed into service. The *Oxford English Dictionary* defines anthropinism as 'consideration of things in their relation to man'.

2. The particular as explicand

It may be best to begin by enlarging on what was said above, namely, that we should be able to conceive how the particular event is understood in terms of general and unchanging meaning. One should notice that the problem of knowledge is here not posed in what is nowadays the most usual way. We have seen that empiricists and positivists try to show how general propositions and their meaning are derived from or judged (verified, tested) by particular experiences. They want to explain the general and for this they take the particular for granted. For some, including probably the positivists in economics, the purely philosophical enterprise of explaining the nature and meaning of general propositions may hold little appeal and it seems that they regard hypotheses (general propositions) merely as a means for predicting particulars or for finding the determinants of particulars. Nevertheless, they want to find and test hypotheses and for this they take particulars for granted in the sense that it must be possible to know (the meaning of) particular events without such hypotheses because they are used to test hypotheses. Here, however, the problem is being posed the other way around, in the way that it was posed by Aristotle and by Kant. It is in the particular event that is to be explained and for this one has to take meaning for granted.

How can this reversal of what to many seems the normal procedure be justified? That the particular should be taken for granted came naturally, as we have seen, to an empiricism which took its bearings from the mechanical universe of matter in motion governed by laws, i.e. from classical mechanics. An impulse is conveyed through space and via the sense organs and nervous system to the brain. There it is decoded into a particular experience and forms the starting point for the explanation of general ideas and meaning. Locke's analysis with its primary and secondary qualities is perhaps the most prominent example of this kind of empiricism. However, what may be called the present-day popular scientific view, incorporating what is generally known of human physiology, the transmission of light, the vibration of air and so on, takes very much the same view. From as long ago as the time of Hume (who heeded Berkeley's critique) positivists have been reluctant to accept this analysis. They did not of course regard it as false – even the simple test of placing one's hands in front of one's eyes can be most convincing. They objected to it because it was extremely theory-laden, because it involved a very sophisticated conception. It presupposed just that kind of general proposition the nature of which they wanted to explain. Unwilling to accord the conception of classical mechanics any special status, they accorded primacy only to sense data, without any presuppositions about how these were acquired. In this approach, however, they accepted uncritically the way the empiricists had posed the problem and, as a result, they faced the problem of finding a theory-free language for expressing sense data. They had to assume that sense data could be articulated into elementary propositions without the intercession of the general propositions which they were meant to explain.

We have seen at length that no one has yet been able to show satisfactorily how this may be done. On the contrary, time and again the conclusion has been reached that there are no theory-free statements of fact, none that are not paradigm-based.[1] It has been a commonplace in the Kantian tradition but has often been reached without explicit acknowledgement to the tradition. Kuhn, for instance, merely discusses a variety of experiments conducted by others and concludes: 'Surveying the rich experimental literature … makes one suspect that something like a paradigm is prerequisite to perception itself.'[2]

If it appears, in the light of the history of the problem of similarity, that particular facts presuppose theories, general propositions or prior conceptions, then it also appears that empiricists and positivists pose their problems the wrong way around. It does not seem sensible to set out to show how particular facts give rise to their own presuppositions or, as in economics, to show how particular events are determined, when this is what one has to know before one can know the particular events. It is surely more sensible to show how the experience of particular events derives its meaning from prior conceptions.

However, a positivist may rightly point out that it is the correctness of conceptions, the validity of hypotheses, that is at issue. If there are no theory-free facts, by what criteria are conceptions judged? This is of course no idle question. Popper seems to have wrestled with it. Polanyi sought the answer in commitment based on a kind of intellectual aesthetics. Some have suggested logical coherence as the criterion, others have put forward a pragmatic criterion. The question is of practical importance in economic life only in cases of dispute about what is the actual state of affairs and how some agreed upon end and may be achieved. In such not uncommon cases, it seems that the most fruitful first step is to lay bare the conceptions or hypotheses to which the parties to the dispute respectively have committed themselves and this may be done by examining particular events as perceived respectively by them. In the first instance at least, the problem is again to show how the experience of particular events derives its meaning from prior conceptions. In what follows, therefore, the particular event will be the explicand. But when particular events are to be explained, meaning (but not conceptions) have to be taken for granted.

3. Meaning and articulation

It may seem absurd to propose to take meaning for granted when so much intellectual effort has been spent on inquiries into meaning. From at least the time of Socrates, the quest for meaning has been considered a commendable task and the linguistic and analytic philosophies of this century have been particularly active in this regard. Taking meaning for granted is certainly not intended to preclude inquiries into meaning and it is therefore necessary to explain what is meant by this proposal.

'Meaning' is an extremely troublesome and confusing term which has to be discussed before one can explain what is meant by taking meaning for granted. The verb *to mean* may simply express intention, as in 'He means well'. Often it expresses the specific intention to indicate or convey, by speech or in some other manner, something one has in mind. 'To mean' is 'to want to indicate'. However, words are also said to mean something and presumably they do not actually want to indicate anything. It is of course the person using words who wants to indicate and does indicate by using words as signs. The meaning of a word, or the meaning that a word has, may therefore be called a significance. As the later Wittgenstein and others were at pains to point out and as is generally known in any case, a word may have a variety of meanings in different contexts. On this basis, the meaning or significance of words in some combinations is simply that which the person using the words as signs means or wants to indicate. On the other hand, one cannot choose words arbitrarily to act as signs. There is the requirement for intersubjectivity that signs be understood by others. The meaning or significance of a word is therefore also a belief held by speakers and listeners that the word is an effective sign, i.e. that it is associated by many with what is meant. To speak of meaning can therefore be confusing because there is (a) meaning as an activity like thinking or at least as an intention to indicate, (b) meaning as that which is meant or indicated and

(c) meaning as an association between sign and signified (without which there would be no signs).[3]

Far more puzzling, however, is the elusiveness of what is in fact meant or indicated. The idea that we are indicating the things we find around us, as we might by pointing a finger, just will not do, especially in economics. Words, of course, only *have* meaning and *are* not meanings in themselves. We mean or indicate something by them. Meaning is also expressed or indicated, almost universally, by metaphor, analogy, imagery, allegory, parable and so on (e.g. economic expansion, buyout expectations, floating exchange rates, a revival of interest). Like words, these have meaning but are not meanings in themselves. We make a point with them; we mean something by them. But just what that something is, apart from being indicated by the term *meaning*, is not at all clear. There may be a clue in the use of the word *sense*. In the previous paragraph one could have said that words may be used in many senses and the word *meaning* in at least three senses. Perhaps when one speaks of a sense of awe, of duty, honour, sorrow, gratitude or security, or speaks of an appreciation of value, beauty or logic or of a consciousness of freedom, one is trying to give expression to the something that is meant. But language is used to indicate something. One cannot state in language what that something actually is; one can only give it a name.

The elusiveness of meaning, of which anyone can convince himself by introspection, lends weight to Polanyi's idea of tacit intelligence. (He was not the first to think along such lines.) It may be instructive to consider meaning in terms of Polanyi's idea of tacit intelligence and articulation. However, what follows will be only in the spirit and not the letter of Polanyi – it is nowhere to be found in his writings. One may say that in meaning or indicating we are pointing beyond the periphery of the formal or articulate. Language and even articulate thought can only indicate because that is our aim in articulation. Suppose A is trying to persuade B to invest in a new project. B shakes his head and says: 'My capital is tied up in a herd of cattle.' A replies: 'Cattle farmers are having a tough time, I know, but only fools are ever locked in. Get out while you can. I can offer you a better investment.' In this verbal exchange a great deal is indicated or, seen the other way around, a great deal is articulated. The words *capital* and *investment* point to a tacit appreciation of much of what is involved in conducting one's affairs in a certain economic environment. If one were to attempt to apply names to it, it would probably include the idea of instrumentality, the sense of security of possession afforded by private property, the sense of insecurity created by competitive conditions as well as all that is implied by the metaphysical use of the word *liquidity* in economics, i.e. the opportunity of changing the form of one's possessions through exchange and so on. The metaphysical uses of *tie up*, *rough time*, *lock in* and *get out* allude to a tacit appreciation of other spheres of life. A's reply not only suggests a whole course of action but in the use of the words *fools* and *while you can* play on certain sentiments. In this way a great deal of tacit meaning is indicated by a few separable sounds. One need not labour the point further. It has often been pointed out that articulation is a considerable human achievement.

However, as the term is used by Polanyi, articulation does not seem to be confined to the case of communication. There is also what one could call private articulation and this form of articulation is of greater interest here. What was, so to say, in the minds of two people in the above example of overt articulation could have been in the mind of a single individual reflecting on the state of his business, perhaps after reading an advertisement inviting subscriptions to a new share issue. In formulating a plan of action or in formulating and attempting to solve a problem, the individual articulates privately. He, so to say, indicates to himself in order to have a clearer grasp of an issue. Articulation disciplines and expands the reasoning powers of man, as Polanyi put it. In the process, tacit and elusive meaning is articulated into entities we may identify with ideas, qualities, forms, universals and so on. Analogous to words, these are signs that have meaning but are not meanings in themselves. Articulate speech indicates to others, articulate thought indicates to oneself. They both point beyond themselves, i.e. beyond the articulate, to tacit meaning, which from now on will be referred to simply as the *intelligible*.

The entities of articulate thought together with certain beliefs, to be dealt with later, will be referred to as *conceptions*. These may be identified with the conceptions which (as argued earlier) are prerequisite to perception. They may also be identified with Kuhnian paradigms. In other words, one may accept the view that most people adopt conceptions as they would learn a language. In these terms, paradigms are the conceptions developed by innovative thinkers whose work caught on. Conceptions may also be regarded as mental constructs, but they are not quite the same as the mental constructs in positivistic analysis. This will become apparent when the relation between conceptions and perception and therefore also between the intelligible and the understanding of particular events is considered in the next few sections.

We are now in a position to understand what is meant by the proposal that meaning be taken for granted. It must be clear by now that one can hope only to paraphrase. On this basis, taking meaning for granted is equivalent to taking it for granted that people do articulate what is intelligible to them. If one starts with the end product of articulation, taking meaning for granted implies that all inquiries into meaning lead and stop at a realization that something is simply intelligible and not amenable to further analysis. In analysing the meaning of a word, idea, model or some experience, one may perhaps reduce it to apparently simpler meanings and perhaps even be able to repeat the process for each of these. But taking meaning for granted implies that this process cannot continue indefinitely; that though one may be able to continue piling up synonyms, metaphors, allegories and examples from experience, one eventually realizes that they are all somehow similar and that, within the limits of one's present mental facilities, one has reached what is simply intelligible. In his quest for standards of conduct, Socrates was never satisfied with mere illustrations and demanded to know the One in the Many, the common theme running through all of them. If one accepts articulation as an act of meaning one must accept also that this may be a search for the inexpressible. This is in fact not far from the conclusions reached by modern linguistic philosophy.[4]

Nevertheless, it is not necessarily a worthless pursuit. It has often been said that mathematics is the prime example (translated into present terminology) of a formal elaboration of the purely intelligible. The idea of a set may be as close as one can get to expressing the intelligible directly, and the system of integers to expressing knowing how to count. While non-Euclidean geometry, for instance, may not be intelligible to most people, it is nevertheless a formal elaboration of what is intelligible according to intelligible logical operations. But mathematics is not the only example. Von Mises held the view, as we have already seen and shall have further occasion to notice, that one can at least reach a contemplation of the One in the Many of economics and that it finds its fullest expression in the idea of action or in the idea of purpose. He saw economics as a formal elaboration of the intelligibility of action or purpose, but one could also say that any analysis of economic concepts or of perceived economic events must eventually, in at least one of its strands, reach a point at which one realizes that the meaning of action, purpose or some synonymous expression is simply intelligible and must be taken for granted.

4. The intelligible, the empirical and the real

On the other hand, since the notion of real counts among the intelligible one cannot simply ignore it. When a person asks himself whether x is real or the statement xyz is true he means something like: is it something outside the self, should one take it seriously, can one rely on it? In the spirit of Polanyi one could say that the notion of real arises out a mixture of particular considerations and religious awe, and there is much in the history of thought to support this view. In this sense both the intelligible and the empirical may be considered real. A lone yachtsman, for instance, may be convinced of the reality of both the salty spray that stings his face on a cold morning and the numbering on the sextant/compass he uses to log his position. So if one likes, one can call both the intelligible and an empirical contact *real*. One can say these are simply two levels of reality. But then again the notion of real surely requires that we can recognize something as unreal. But that is not hard to find. If, for instance, our yachtsman were to pass the time by imagining himself to be a Roman notable attended by pretty slave girls, he will if he is sane realize that that is not real. What is missing is a realization, an awareness which has the connotations one could call an empirical content, in other words, intelligible is not simply anything that can be thought of. Even the daydream has meaning, but they are meanings in a certain combination with what we have said is the empirical. This the imagined here and now, particular or there and then, imagined empirical content without the actual awareness of it is an example of the unreal. The opposition of real which gives the real its meaning is the combination of meaning without empirical contact. But fantasy is not the only example of it. Memory, expectations, conjecture are others but there is then some connection with the here and now and we cannot yet deal with them.

Experience in the sense of something being present here and now is a notion which presumably is familiar to everyone. Nevertheless, as the discussion of sense data, elementary propositions, etc. in the previous chapter has indicated, it is by no means easily analysed. The difficulty seems to lie in the contrast that is implied when experience is described as a source of knowledge or as point of contact with the real world. If experience is taken to include the experience of one's own feelings and thoughts (as in the expressions 'a feeling of remorse came over me', 'it occurred to me' or 'the thought struck me'), then it would seem to be coextensive with being conscious of something with all there is for a human being, so that it seems strange to refer to it as a source or a point of contact. A similar though more transparent puzzle arises when one says that it is always today and tomorrow never comes, for, while it is time that we always live today, this obviously makes sense only on the understanding that there were and will be other days. In a similar way, the notion of experience makes sense only in the contrast with a wider-ranging mental activity or on the understanding that it is in the nature of mental activity to transcend what is present here and now.

5. The here and now

As one walks around a building, its appearance continually changes. It is a matter of perceptive or, quite literally, of different points of view. Moreover, the building necessarily appears to one in a particular perspective. One cannot even imagine what it would look like if it were viewed from nowhere in particular. The way things look, sound or feel is not independent of one's own presence and all sense perception is correlative with at least a potential awareness of here and now.

Ideas and concepts, on the other hand, need not be bound to the here and now. We can speak quite generally of value and welfare, choice and decision, saving and investment, as we can of number, virtue or justice, without the where and when of our speaking making any difference to what is spoken of. Ideas and concepts, one could say, are not in time and space. But even things which are in time and space – events and situations, objects and people – may be treated quite generally. Something which has been observed by many is regarded, by virtue of this intersubjectivity, as objectively real. As such it is treated as though there is no need to specify who has seen it or from where and when it has been seen. It is put, so to say, into an impersonal perceptive, or is viewed from a transcendent point of view. Though we cannot imagine what a building would look like if viewed from nowhere in particular, we can nevertheless refer to it without giving any thought whatever to our own presence.

The expression *here and now* in this context is not equivalent to a point (or area and interval) in space and time, or in space and time. The latter is the appropriate terminology when we adopt the transcendent viewpoint mentioned above. We may refer to the points A and B on a diagram or (as time points) in the course of a process without in any way considering how we ourselves are related to the origin of the frame of reference used. But when we say that A is here or now and B is there or then, or that B is further from here or now than A, we are in a sense using our own consciousness as the origin of a frame of reference; we have a particular point of view.

The idea of here and now provides a criterion for distinguishing between the particular and the general and, since the distinction is central to the theme of this study, it will be an essential feature of the proposed concept of the intelligible account that such accounts are drawn up in the here and now. In conventional discourse there is no single criterion for distinguishing the particular from the general. Such discourse is partly based on the species/genus idea and the successively high degrees of abstraction of Aristotelian logic, and this makes the distinction largely a relative one. We may agree, for instance, that a model of buyers and sellers establishing prices in a market is quite general. A model of buyers and sellers establishing prices on stock exchanges is more particular. But is it particular? Would it be particular if a stock exchange were specified by name? On the criterion to be established here, it would not. *Buyers*, *sellers* and *prices* are (general) concepts and the named stock exchange is put into what above was called the impersonal perspective. The particular, as it is to be understood here, must be in a personal perspective. The proposed criterion is a variant of the ostensive test (that one can point to the particular) also used in conventional discourse, namely, *the particular is relative to the here and now of a perceiver*. On this criterion, a particular event or situation can be known to only one individual. All public or *objective* knowledge is general. Furthermore, while there may be degrees of generalities, there can be no degree of particularity.

It is also within the logic of the conceptual scheme proposed here that it is in the here and now, and only in the here and now, that an individual has an empirical content with the so-called external world, with something he believes to have an existence outside of himself. General empirical knowledge can therefore be derived only from the particular, which, as argued above, is the private experience of an individual. (However, this does not mean that existence or reality can be ascribed only to the particular.) An individual can also remember or imagine a particular event or situation. A remembered or imagined particular is relative to a remembered or imagined here and now. The latter differs from *the* here and now in that there is no empirical contact. Empirical knowledge cannot, of course, be derived from an imagined particular.

The idea of the here and now is similar to Shackle's idea of the solitary present, the solitary moment or the moment in being, but he uses it for different purposes, namely, to draw conclusions about the nature of time and decision. 'The time of actuality or consciousness in which perceptions, thoughts, and emotions occur in a unique moment unaccompanied by any other moment, the sole possessor of all that is, the solitary present. … There is but one moment, though it exists by changing.' Since the solitary moment is the only actual time, the past and future exist only as memory and expectations, i.e. as thoughts. Shackle asserts that this 'has consequences which social sciences have often overlooked'. Comparison of actual experiences and different moments would be possible only in an 'extended time seen from outside by an extra-temporal observer' ('who has no business whatever to pronounce on a question in which he has no competence'). The notion of extended time is 'the notion of an endless sequence of moments constituting, as it were, an axis or a one-dimensional space'. It is however, a mental construct; the actual time of the solitary moment is 'a forceps which grips us between the past which is unchoosable and the future which is unknowable'.

Though these ideas are somewhat novel within economics, similar ideas are very old in philosophy. For instance, Aristotle introduced his discussion of time by questioning whether time may be said to exist at all. 'One part of it has been and is not, while the other is going to be and is not yet.' There is a now, but time is not made up of nows. Time, according to Aristotle, is a measure of motion or of change. When we judge the now to be different (from what we remember), we comprehend a before and an after a change and take it that time has elapsed in between. In another sense, however, the now is always the same. It is always 'the end of that which is past and the beginning of that which is to come'.

6. Intelligibility

The influence of physical science on economics is epitomized in the way the function of theory is most commonly perceived. The function of theory is not only to formulate coherent conceptual systems but to do so in order to represent the world as a determinate system. It is as though anything else is inconceivable. What other function could theory possibly serve if not to find determinants? What else could explanation possibly mean? How else could theory be of use to anyone? Determinate systems are not merely the subject matter of the physical sciences, the special preserve of these sciences, but the sum and substance of all science.

It is perhaps on account of this perception that the exponents of a subjectivist approach to economics have had a rather moderate influence on their colleagues. What they have had to say has no doubt been found interesting and relevant to economic questions understood in the everyday sense. But what can be done about their observations? They seem to defy almost all attempts at incorporation into determinate systems. They belong, as we have put it, to another domain of thought, to a different side of experience. The role of expectations in economic affairs is a good example of this. That decisions are based more on what is expected to happen than on what has already happened seems to be a reasonable proposition. Mainstream economic theorists have therefore bent over backwards to find some way of incorporating expectations into their representation of economies as determinate systems subject possibly to stochastic shocks. But did those who drew attention to the importance of expectations really regard this transfer from the subjective to the physical domain of thought as a prerequisite for the theoretical treatment of expectations? Writing on expectations in economics as a social science, Lachmann remarked in an early paper that ‘it is intelligibility and not determinateness that social science should strive to achieve’.[5] In a later publication he said: ‘In social theory our main task is to explain observable social phenomena by reducing them to the individual plans … that typically give rise to them’. This is what Weber meant by the explanation of action ‘in terms of the meaning attached to it by the actor’ and

> Human action is not *determinate* in any sense akin to the one in which natural science has to strive for the *determinacy* of the events it studies. A mechanistic interpretation of action, couched, say, in terms of *response to stimulus*, would have to explain away such simple facts as that different men in identical situations may act differently because of their different expectations of the future.[6]

Our final task will be to suggest that the study of economic ideals, such as the ideals of market order, not only becomes feasible but gives us much scope for useful work when we try to achieve intelligibility rather than determinateness.

Inquiry in which intelligibility rather than finding determinants is the basic concern may take on a variety of forms. History is probably the best-known example of such inquiry (except of course when it is informed by theories of historical determination). Historical inquiry seeks intelligible accounts of how one thing led to another, and of the mental life that was involved, without implying that the course of events could not possibly have been otherwise, i.e. without implying that it was determined in the physical sense. Max Weber was a pioneer in the field of adapting this method of interpretation to the social sciences. Theory then has the role of providing conceptual schemes and conceptual clarification.[7] Menger also seems to have seen the role of his economic theory in this way.[8] Mises coined the term praxeology for the study of the implications of the concept of human action. Economic theory, according to Mises, is one form of praxeology.[9] Hayek speaks of a compositive (as opposed to an analytic) method. From simple features of mental life familiar to us all, compositive theory infers more complex social phenomena in very general terms, rather as a conjectured history of money has been constructed. He considers this to be the correct method for understanding the unintended consequences of human action, i.e. for understanding spontaneous order, and it is, according to Hayek, the method of economic theory. He regards general equilibrium theory as a compositive theory explaining the general interrelations of prices as an unintended consequence of purposeful action.[10] The word *consequence* has quite different connotations in the languages of physical science and of history respectively and it is not entirely clear which ones Hayek had in mind when he spoke of general equilibrium and the compositive method.

These forms of inquiry have in common a concern with the mental life of individuals in relation to their circumstances. The aspects of this relationship that usually come under consideration may be divided roughly into three kinds. First, there are the physical conditions that are obstacles to the attainment of human ends, but which may also be exploited with the aid of technology to serve human ends. Secondly, there is the fact that individuals share a common environment and that their individual endeavours to achieve specific ends therefore impinge on each other. Thirdly, there are the ideals, values and moral precepts that are held. They create obligations which may modify the ends individuals seek but they also create opportunities for each individual to exploit the obligations of others to serve his own ends. Above all, they may be mutually contradictory and that creates problems in a one-world shared environment. Economic studies have usually emphasized only the first two aspects. It is, however, the third aspect which seems to have the potential for making economic studies empirically more specific, though considerable conceptual clarification would be needed. We shall argue this case first. Thereafter we shall try to give some indications of how such empirically more detailed analysis may enhance our understanding of issues of long-standing economic interest.

7. Determinism, intelligibility

According to Shackle, Keynes' 'ultimate thesis' was that economic actions, in particular investment and hence output and employment, 'rest upon the most mutable and elusive of all economic elements', namely, expectations and confidence which may be formed or dissolved by a mere 'breath of suggestion from *the news*'. Shackle asks: 'Where, in such a vision, is the basis of mechanistic, or hydraulic, determinism, which 'for orthodox analysts was the presupposition and *sine qua non* of analysis?' This section considers how we are to understand the determinism referred to here, which, as Shackle suggests, is incompatible with familiar, everyday views of economic affairs. It also suggests a way in which non-deterministic economic theory could be formulated.

Classical mechanics is the paradigm of a deterministic theory. It has a vague affinity only with the limited and not very successful area of dynamic economic theory and the related stability conditions of equilibrium – an affinity which Samuelson pointed out, with some pride. In static general equilibrium analysis, the word *determinate* refers to the circumstance that a set of simultaneous equations has at least one solution and the solving of such equations is often taken to represent a hard-to-visualize process of mutual determination. But the determinism which Shackle argues against is a much wider and more vague notion and carries with it the implication that the entire course of history is predetermined and inevitable. As expressed in a well-known passage by Laplace, it appears to have arisen as a logically unwarranted though understandable extension of classical mechanics and is often seen as the guiding light for all scientific endeavour. A strong case can be made out that such determinism is incompatible with our notion of choice ('can choice be so-called, if it is *determinate?*' asks Shackle) and with our opinions on how we come by the expectations and worldview that influence our economic actions; and further that it is deterministic theory which should give way in order to remove the incompatibility.

The question then arises what there is to say in economics if the economic theorist cannot take it for granted that his task is to show the determinants of that which he wishes to explain. An answer which has sometimes been proffered is that the economist can help to make the past intelligible. But this has difficulties of its own and, while it may be a very worthwhile pursuit, it is perhaps not one to the liking of most economists. Another approach is then outlined, one which may be illustrated by the long association between economic theory and arguments in favour of free markets. Adam Smith could make out his case without implying a universal determinism and so could Walras if general equilibrium analysis is used in the way in which he apparently intended it to be used. (It was Pareto who put a deterministic interpretation on the theory.) Economic theory in this application is rather like a disguised definition of a market which may be used as a principle of, or a guide to, empirical investigation. The notion of equilibrium itself, von Mises' praxeology and even Ricardo's analysis of the stationary state may be seen in the same light, i.e. as definitions to guide investigation. However, when economic theory is regarded in this non-deterministic way, new problems loom on the horizon. These would appear to require nothing less than a new formulation of logic appropriate to the economist's sphere of interest.

8. Articulating subjectivist economics: from ontology to codification

a. Distinguishing ontological subjectivism from the logic of subjectivism

It would be useful to distinguish between ontological subjectivism and a logic of subjectivism. Ontological subjectivism is the commitment to a subjectivist *Weltanschauung* or a subjectivist ontology, i.e. a set of beliefs or tenets. (This is the current usage of *ontology* which Quine appears to have popularized.) Different people may of course be committed to different subjectivist ontologies. They may also be committed to mechanistic or physicalist ontologies. Ontological subjectivism is reflected in the unself-conscious natural attitude to life. We know that there are people out there, that they have some knowledge, beliefs and expectations, that they make decisions, make plans, act purposefully and so on. Ontological commitment is necessary in any field of study but does not in itself provide a method of analysing what one is committed to.

We may speak of a logic of subjectivism which is a logic in the sense that it deals with the language in which subjectivist ontologies are expressed. One speaks about the terms in which one speaks about people, economic affairs and so on. As philosophers put it, one uses a metalanguage to talk about or mention the terms which one uses in an object language. For example, let us consider the statement 'Economic phenomena must have an identifiable source in some mind or be generated by human action.' One may make much the same point, and incidentally avoid, the obvious mechanical metaphors by using a metalanguage thus. '*Economic phenomena* are properly expressed or defined in the terms in which *mental attitudes* and *human action* are expressed.' The words in italics are in the object language and the rest is in the metalanguage. One is no longer speaking about economic phenomena but about the expression *economic phenomena*.

When a writer's intention is not only to state but also to analyse his subjectivist ontology, he ventures into the logic of subjectivism. But he may do so without using the distinction between metalanguage and object language, making use instead of metaphors and possibly other devices in what would be the object language. The original statement about economic phenomena above may be a case in point.

Let us consider, by way of example, one of Shackle's interesting discussions of *decision*, where he seems to waver between discussing the activity of making decisions and the usage of the term *decision*. He says that in a determinist world, in which 'history is a book already written, whose pages the hand of time is merely turning, not composing', human beings 'only seem to themselves to be the source of any current of events' and decision is illusory. In a world of certainty, of perfect foresight, decision is empty since it is really calculation. In 'a world without discernible order' where one has no idea what the sequels of one's actions will be, decision is powerless. He went on to ask: 'What kind of world can we find, wherein decision will be none of these things?'[11] Shackle uses metaphors effectively to talk about the activity of making decisions, but he is really concerned with the usage of the term *decision*. He also poses the problem in a more metalinguistic way:

> How to find a scheme of thought about the basic nature of human affairs, which will include *decision* in the meaning we give to this word in our unself-conscious, intuitive attitude to life, where … we take it for granted that a responsibility lies upon us for our acts; that these acts are … *creative, inceptive,* the source of *historical novelty*.[12]

To look for a scheme of thought is, in our metalanguage, to inquire into which terms hang together and how they are interlinked. (Metalanguages also use metaphors.) In a way, this is what Shackle has been doing in most of his work. But it is a voluminous work. What is admirable about the physical sciences, and mechanics in particular, is that they have formulated their domains of thought succinctly. Shackle is able to say that decision in a determinist world 'belongs to a conceptual world where morality, ethics, wisdom and, above all, creative thought have no place',[13] because it is quite clear what does have a place in the conceptual world or domain of thought of mechanics. There is evidence in the Menger-Mises tradition in the Austrian school of a groping for the same clarity in the subjectivist domain of thought. To this we now turn.

b. Implicit definitions and codifying domains of thought

There has been much argumentation over whether Newton's famous three laws of motion are concealed or implicit definitions.[14] It is unlikely that Newton regarded them as such, but they may be interpreted as statements which show how certain terms are interlinked and which helped in this way to articulate and codify the pre-existing thought domain of mechanics.

Let us consider the law of inertia. Everybody perseveres in its state of rest, or of uniform motion in a right line, unless it is compelled to change that state by forces impressed thereon.[15] This original formulation by Newton is full of anthropomorphic metaphors: *persevere*, *rest*, *compel* and even *force* with its old usage in the context of human effort and muscular exertion. But, while the metaphors are suggestive, the main terms in the law of inertia mutually define each other and nothing more than their stated mutual interconnection may be deduced from any one of them. It has sometimes been said that the law of inertia states that a body continues to move in a straight line at a constant velocity except when it does not. When it does not, one says that forces are acting upon it. In the thought domain of classical mechanics, force is a measure (as implied by the second law of motion) of change of velocity or direction and no more than that. The limited meaning of straight *line* or of *uniform motion* are derived similarly. When the latter is elaborated into *equal distances in equal time intervals* the meaning of these terms (e.g. equal time intervals) in mechanics may also be derived. How the law of inertia may be used in practical applications is not our concern here. What interests us is the way it shows what belongs and what does not belong to the thought domain of mechanics and how the terms are interlinked. One may call the law of inertia a part of a codification of the thought domain of mechanics. One could also refer to a paradigm or a Wittgensteinian language game in the sense of a stipulation of the usage of terms.[16]

Though it is a tall order, it would be useful to develop a codification of the subjectivist domain of thought, the subjectivist language game, which would, for instance, stipulate the usage of Shackle's *solitary present* or *time lived in* and of *decision* in the required sense. It would not show us what people and the world are really like, but by more clearly delineating the implications of what we are committed to ontologically, it may make decisions in this regard easier. The current discussion of uncertainty, for instance, seems to involve (though one cannot be sure of this) quite different domains of thought when some economists treat uncertainty as a feature of the environment and others as a mental state of those who have doubts about the environment.

It is common in economic writings to provide ad hoc clarifications of concepts by such devices as the Crusoe economy or Adam Smith's 'early and rude state of society'.[17] Menger and Mises, however, are rather special cases. Though their respective philosophical tenets made them express themselves differently, they may be seen to have advocated a systematic codification of the thought domain of economics, to have held it up as the ideal form of economic theory.

In the preface to the *Grundsätze*, Menger touched on mechanomorphism when he spoke of an empty playing with superficial analogies between the phenomena of economic life and of nature. Yet he insisted that his method was the same as that which had led to great success in the natural sciences.[18] His position became clearer in the *Untersuchungen* when he explained what he meant by *exact science*.[19] Still later he said that sciences differ by subject matter but that a certain direction of cognitive endeavour

(*Richtung des Erkenntnisstrebens*) or way of looking at things (*Betrachtungsweise*) establishes exact branches in each and allows, for instance, the formulation of the morphology of economic phenomena.[20] Though space does not permit it here, it may be shown that when Menger is translated out of the Aristotelian into a modern style of philosophizing, his phenomenal forms (*Erscheinungsformen*) become something like terms and the exact laws based on them something like a codification of the economists' language game, or, as we may say, of the thought domain of economics. For instance, he spoke of exact laws in economics as laws of economizing (*Wirtschaftlichkeit*).[21] Then in a book review of 1887 he described them as laws of rational economic means-ends relations (*Gesetze der rationalen ökonomischen Zweckbeziehungen*).[22]

Mises, it would appear, translated Menger into a neo-Kantian and later an ever more radically rationalist framework and in this way was led to praxeology. Trimmed of its claims of apodictic certainty and incontestability, the category of action and hence praxeology translates into our metalanguage as the thought domain of human action. Perhaps because he conceived action as a Kantian category which all of us carry around with us all the time, he nowhere, as far as one can tell, tried to set out in a few propositions what praxeology entails apart from means and ends, i.e. he made no attempt to codify it. His discussion, in one of his books, on the logical character of praxeology, was over and done with in about 300 words.[23] This is a pity. Even if we do all have a good idea of what it is to act, it is not easy to articulate the further reaches of the implications of that idea. Mises said, for instance, that the 'category *means and ends* presupposes the category *cause and effect*' and even that 'causality is a category of action'.[24] But, as is evident from Shackle's problem with *decision*, it is not so obvious how the troublesome idea of causation fits in with the idea of human action. What of private property? It seems somehow to imply means and ends, but not vice versa. Perhaps all the material for a codification of praxeology may be found in *Human Action* if someone is prepared to go through it with a fine-tooth comb. There have of course been excellent commentaries on praxeology.[25] But something else is needed: a few succinct propositions in which the main terms are intended to be mutually defined, not with the arbitrariness of some axiomatic systems, but so that at least commonly found ways of thinking are reflected. It would by no means be an easy task.

c. The coherence rule: maintaining consistency within thought domains

Let us imagine that we have available to us codifications of various domains of thought or, more likely, competing versions about which argument continues. We would then be in a position to suggest what we may call the *coherence rule*. The rule stipulates that a question should be posed in terms which all belong to the same domain of thought and that the corresponding answer should be composed in terms which all belong to the same domain of thought as the terms of the question.

With the aid of the coherence rule, the logic of subjectivism would be able to reformulate the case which ontological subjectivism brings against mechanomorphism. This case usually has the form: The real world is not like that. The *that* which by implication is unreal or nonreal is a proposition of a theory, conjecture, intended representation or anything which (as the metaphor has it) exists only in the mind. It is intended as an explanation or a solution to a problem or a representation for which one of these is the ultimate goal. Loosely it may be called an answer. To every answer there must correspond a question. When the *real world* is brought into relation with theory (i.e. answers in our loose terminology) in the case against mechanomorphism, we must mean some question, problem or issue in the real world. In any case, we could not be concerned with a description of the world as a whole; the real world is a concept without bounds. With the real world as question and theory as proposed answer, the coherence rule may be applied. The case against mechanomorphism is that the coherence rule should be applied.

The remarkable success of classical mechanics and of the physical sciences in general may owe much to the fact that something like the coherence rule is followed. However complex and full of life a situation may be, one is able to pick out the mechanical aspects (if there are any) and explain them entirely in mechanical terms. One does not go on a detour through thought domains where, say, gremlins, fate or the forces of darkness are at home, nor does one end up in the subjectivist domain of thought. Similarly, when an issue is expressed entirely within the subjectivist domain of thought, one should ensure that its explanation remains in that domain and does not end up in the thought domain of mechanics.

This is the strictest form of coherence rule. There may be less restrictive versions. Let us imagine a street map which is fixed to a wall and has on it an arrow with a notice stating: *You are here*. It is a device which establishes a connection between a here and now and a point in a representation of mechanical space. If here and now is something like Shackle's time lived in, it establishes a connection between the thought domain of Shacklean mental life and that of mechanics. This may require a reformulation of thought domains or allow a relaxation of the coherence rule. It is hard to tell.

More broadly, however, the coherence rule may be regarded as unduly restrictive. Some writers on economic methodology have lately been taken with the notion of pluralism.[26] They see it, it appears, as a welcome liberation from positivism and falsificationism. But it has the potential for creating a free for all in which no intellectual rules apply. In this spirit we may well ask why we should lumber ourselves with a rule. Indeed, there is no profound reason in the grand design of things why we should apply the coherence rule, or why we should not seek our explanations by whatever means we can think of.

The coherence rule is an ideal. It is a version of the ideal which has come down to us from ancient Greece, not of a simple monism, but of distinct but internally coherent intellectual disciplines. Let us take an example. On separate occasions we may say: *leave the decision to the market*; *the new regulation has paralysed the market*; and *the market is operating smoothly*. On their respective occasions, each of these statements is no doubt quite meaningful. But if someone were asked to develop an intellectual discipline called economics in which the three statements may be shown to be part of a coherent conception, he would not know whether the market should be treated as a mind capable of making decisions, as a biological organism or as a machine. The metaphors, however, need be no more confusing than those in Newton's original formulation of the law of inertia if they are tied together by mutual definition into a single domain of thought. The coherence rule may then be applied.

What is disconcerting about textbook expositions of economics, at least to one who is not impressed by the idea of testable hypotheses, is that the coherence rule is not followed. The student is introduced to a topic reeking with the richness of social life. He is then taken by a little legerdemain through a blur and suddenly finds himself in an eerie world of continuous functions. He watches the functions shift about and, when they have stopped, notes down the coordinates of their points of intersection. He is then taken again through the blur and, behold, he finds himself once more among familiar human faces. The recommendation of this study is that the subjectivist case against mechanomorphism be based on the ideal that such blurs be removed.

Notes

1 Hayek has intimated that he borrowed the term 'rule of conduct' from David Hume.

2 Ryle (1949, pp 25–61).

3 Polanyi (1958, Part 2, pp 69–245).

4 'Rules, Perception and Intelligibility' in Hayek (1967, pp 43–65; see especially pp 45–46).

5 Lachmann (1943b, p 14).

6 Lachmann (1970, pp 31 and 36).

7 See Lachmann (1970, 'The Method of Interpretation', pp 17–48). The denotation of the word 'hermeneutics' has recently been extended to cover this method.

8 In the case of Menger the matter is bedevilled by the fact that his Aristotelian presuppositions did not allow him to speak of conceptual schemes but rather of phenomenal forms (*Erscheinungsformen*). But these amount to concepts. In the Foreword to the *Grundsätze*, for instance, he said that reference to human free will may well be used to contest the 'Gesetzmässigkeit' (which may have a meaning of conformity to theoretical principles) of economic action ('wirtshaftliche Handlungen'), but never of the phenomena. It is the latter which are the objects of inquiry in economic theory (Menger 1871, p xlvii). He dedicated the *Grundsätze* to Roscher and ended the Foreword with a friendly greeting

from Austria to the German historical economists to whose movement, he said, his attempted reform of the principles of economics belonged and on whose work it had been based (Menger 1871, pp xli and xlviii). His later altercations with these economists removed the friendliness but not the point he was making. In the *Untersuchungen* he said that historical inquiry was conducted by means of an understanding of phenomenal forms and even justified his position with a reference to Adam Smith's 'The History of Astronomy' (Menger 1883, pp 18–24). In this he was mistaken because Smith's stance there is that of David Hume rather than of Menger's version of Aristotle.

9 Mises (1949) is really a long elaboration of this point.

10 Hayek (1952, Chapter IV). The point about general equilibrium is made on p 43. See also Hayek (1967, 'The Theory of Complex Phenomena', pp 22–42).

11 Shackle (1966, pp 72–74).

12 Shackle (1966, p 73).

13 Shackle (1966, p 72).

14 For a discussion of this question, see Nagel (1979, pp 174–202).

15 Quoted in Nagel (1979, p 158). The law of inertia is the first law or axiom of motion.

16 When Kuhn introduced the term *paradigm* in the sense now common, he mentioned Newton's *Principia* as one of the works which implicitly defined problems and methods for succeeding generations of scientists. When he dissected the term in his *Postscript 1969*, he used Newton's second law ($f = ma$) to illustrate paradigms as implicit definitions and as group-licensed ways of seeing situations as like each other. The latter could be taken to refer more generally to the ability to use metaphors effectively. See Kuhn (1970 [1962], pp 10, 182–184 and 188–191).

17 Smith, A. (1937 [1776]) *The Wealth of Nations*, Book I, Chapter VI, first sentence. The drawback of ad hoc clarifications is illustrated by the fact that the beaver-and-deer case, which is introduced with these words, has often been taken, perhaps even by Ricardo, as the basis of a labour-content theory of value. It can also be understood to introduce the simple idea that prices resolve themselves into factor earnings.

18 Menger (1871, Vol I, p xlv).

19 Menger (1833, First Book, Chapters 4 and 5, pp 31–59).

20 Menger (1889, Vol III, pp 189–192 and 197–198).

21 For example, in Menger (1883, pp 59 and 265).

22 Menger (1933–6, Vol III, p 105). See also p 192 where he speaks of an understanding of the inner connection (*inneren Zusammen-hanges*) of the results of scientific research.

23 Mises (1978, pp 44–45).

24 Mises (1949, pp 22–23).

25 Lachmann (1951); Lachmann (1977, pp 94–111); Rothbard (1976). Rothbard, among others, explores praxeology as a logic.

26 As opposed to monism, i.e. the assumption that there is a single ultimate principle or kind of entity. See Caldwell (1982, Chapter 13, pp 244–252); Boland (1982, Part IV, especially Chapter 12, pp 188–196).

Select Bibliography

Aaron, R.I. (1967) *The Theory of Universals*, 2nd edn, Oxford: Oxford University Press.

Alter, M. (1982) 'Carl Menger and *Homo Oeconomics:* Some Thoughts on Austrian Theory and Methodology', *Journal of Economic Issues* 16: 149–160.

Aristotle (1925 [350 BCE]) *Nichomachean Ethics*, translated by W.D. Ross, Oxford: Clarendon Press.

Ayer, A.J. (1940) *The Foundations of Empirical Knowledge*, London: Macmillan.

Ayer, A.J. (1956) *The Problem of Knowledge*, Harmondsworth: Penguin Books.

Ayer, A.J. (1978) *The Listener* 99(2549): 270.

Berkeley, G. (1901) *The Works of George Berkeley*, edited by A. Campbell Fraser, Oxford: Clarendon.

Bentham, J. (1843) *The Works of Jeremy Bentham*, Edinburgh: William Tait.

Black, R.D.C. (1973) 'W.S. Jevons and the Foundation of Modern Economics', in R.D.C. Black et al (eds), *The Marginal Revolution in Economics*, Durham: Duke University Press, pp 98–112.

Black, R.D.C. et al (eds) (1973) *The Marginal Revolution in Economics*, Durham: Duke University Press.

Blaug, M. (1968) *Economic Theory in Retrospect*, 2nd edn, New York: Richard D. Irwin Inc.

Boehner, P. (1957) *Ockham: Philosophical Writings*, Edinburgh: Nelson.

Boland, L.A. (1979) 'A Critique of Friedman's Critics', *Journal of Economic Literature* 17(2): 503–522.

Boland, L.A. (1982) *The Foundations of Economic Method*, London: Allen & Unwin.

Borch, K. (1973) 'The Place of Uncertainty in the Theories of the Austrian School', in J.R. Hicks and W. Weber (eds), *Carl Menger and the Austrian School of Economics*, Oxford: Clarendon, pp 61–74.

Bradley, F.H. (1893) *Appearance and Reality: A Metaphysical Essay*, London: Swan Sonnenschein & Co.

Broekman, J.M. (1963) *Phänomenologie und Egologie: Faktisches und Transzendentales Ego bei Edmund Husserl*, Den Haag: Springer.

Cairnes, J.E. (1874) *Some Leading Principles of Political Economy*, London: Macmillan.

Caldwell, B. (1982) *Beyond Positivism*, London: Allen & Unwin.

Carnap, R. (1967) *The Logical Structure of the World*, London: Routledge (translated from *Der Logische Aufbau der Welt* [1928]).

Cassirer, E. (1953 [1910]) *Substance and Function*, Dover: Dover Publications.

Cherniss, H. (1955) *Aristotle's Criticism of Presocratic Philosophy*, Baltimore: Johns Hopkins University Press.

Clerc, J.O. (1942) 'Walras and Pareto: Their Approach to Applied Economics and Social Economics', *The Canadian Journal of Economics and Political Science / Revue Canadienne d'Economique et de Science Politique* 8(4): 584–594.

Cohen, M.R. and Nagel, E. (1934) *An Introduction to Logic and Scientific Method*, London: Routledge.

Collingwood, R.G. (1945) *The Idea of Nature*, Oxford: Clarendon.

Dewey, J. (1938) *Experience and Education*, New York: Macmillan Company.

Dolan, E.G. (1976) *The Foundations of Modern Austrian Economics*, Kansas City: Sheed & Ward.

Eddington, A. (1939) *The Philosophy of Physical Science*, Cambridge: Cambridge University Press.

Elliston, F.A. and McCormick, P. (1977) *Husserl: Expositions and Appraisals*, Notre Dame: University of Notre Dame Press.

Ewing, A.C. (1961) *The Fundamental Questions of Philosophy*, London: Routledge & Kegan Paul Ltd.

Fair, R.C. (1978) 'A Theory of Extramarital Affairs', *Journal of Political Economy*, 86(1): 45–61.

Feyerabend, P.K. (1981) 'An Attempt at a Realistic Interpretation of Experience', in *Realism, Rationalism and Scientific Method. Philosophical Papers*, Vol 1, Cambridge: Cambridge University Press, pp 17–36.

Fisher, I. (1918) 'Is "Utility" the Most Suitable Term for the Concept It is Used to Denote?', *American Economic Review*, 8: 335–337.

Freeman, K. (1953) *The Presocratic Philosophers*, 3rd edn, Oxford: Blackwell.

Frege, G. (1894/1977) 'Review of Dr. E. Husserl's Philosophy of Arithmetic'. First published 1894 in J.N. Mohanty (ed) *Readings on Husserl's 'Logical Investigations'*, The Hague: Martinus Nijhoff, pp 6–21.

Friedman, M. (1953) 'The Methodology of Positive Economics', in *Essays in Positive Economics*, Chicago: University of Chicago Press, pp 3–43.

Gellner, E. (1974) *Legitimation of Belief*, Cambridge: Cambridge University Press.

Guthrie, W. (1962) *A History of Greek Philosophy*, Vol I, Cambridge: Cambridge University Press.

Hayek, F.A. (1949) *Individualism and Economic Order*, London: Routledge.

Hayek, F.A. (1952) *The Counter-Revolution of Science*, Glencoe: Free Press.

Hayek, F.A. (1967) *Studies in Philosophy, Politics and Economics*, London: Routledge.

Hicks, J.R. and Weber, W. (eds) (1973) *Carl Menger and the Austrian School of Economics*, Oxford: Clarendon.

Hollis, M. and Nell, E. (1975) *Rational Economic Man*, Cambridge: Cambridge University Press.

Howey, R.S. (1960) *The Rise of the Marginal Utility School 1870–1889*, Lawrence: University of Kansas Press.

Hume, D.A. (1962 [1739]) *Treatise of Human Nature* (Books I and II first published 1739; Book I, Part III, Section V edited by D. Macnabb), London: Fontana Library.

Hutchison, T.W. (1938) *The Significance and Basic Postulates of Economic Theory*, London: Macmillan & Co.

Hutchison, T.W. (1953) *A Review of Economic Doctrines 1870–1929*, Oxford: Clarendon.

Hutchison T.W (1960) 'Preface', in *The Significance and Basic Postulates of Economic Theory*, 2nd edn, New York: Augustus M. Kelley.

Hutchison, T.W. (1973) 'Some Themes from Investigations into Method', in J.R. Hicks and W. Weber (eds), *Carl Menger and the Austrian School of Economics*, Oxford: Clarendon.

Jaffé, W. (1935) 'Unpublished Papers and Letters of Leon Walras', *The Journal of Political Economy* 43: 87–207.

Jaffé, W. (1973) 'Leon Walras's Role in the "Marginal Revolution" of the 1870s', in R.D.C. Black et al (eds), *The Marginal Revolution in Economics*, Durham: Duke University Press, pp 113–139.

Jaffé, W. (1976) 'Menger, Jevons and Walras De-Homogenized', *Economic Inquiry*, 14(4): 511–524.

James, W. (1931 [1907]) *Pragmatism*, New York: Longmans.

James, W. (1950 [1890]) *The Principles of Psychology*, Mineola: Dover Publications.

Jevons, W.S. (1874) *Principles of Science*, London: Macmillan & Co.

Jevons, W.S. (1879a) 'Methods of Social Reform, II: A State Parcel Post', *The Contemporary Review*, XXXIV.

Jevons, W.S. (1879b) 'The Periodicity of Commercial Crises, and its Physical Explanation', *Journal of the Statistical and Social Inquiry Society of Ireland* VII(54): 334–342.

Jevons, W.S. (1957 [1871]) *The Theory of Political Economy*, 5th edn, New York: Kelley & Hillman.

Kant, I. (1793) *Critique of Pure Reason* (first published 1781, revised 1787).

Kauder, E. (1965) *A History of Marginal Utility Theory*, Princeton: Princeton University Press.

Keynes, J.M. (1936) *The General Theory of Employment, Interest and Money*, London: Macmillan.

Keynes, J.N. (1917) *The Scope and Method of Political Economy*, 4th edn, London: Macmillan.

Kirk, G.S. (1954) *Heraclitus: The Cosmic Fragments*, Cambridge: Cambridge University Press.

Kirk, G.S. and Raven, J.E. (1957) *The Presocratic Philosophers*, Cambridge: Cambridge University Press.

Kirzner, I.M. (1979) *Perception, Opportunity and Profit*, Chicago: University of Chicago Press.

Knight, F. (1940) 'What is Truth in Economics', *Journal of Political Economy* 48: 1–32.

Kuhn, T.S. (1970 [1962]) *The Structure of Scientific Revolutions*, 2nd edn, Chicago: University of Chicago Press.

Labia, J. (1988) 'Review Article on McCloskey's book *The Rhetoric of Economics*', *South African Journal of Economics*.

Lachmann, L.M. (1943a) '"Elastic Expectations" and the Austrian Theory of the Trade Cycle', *Economica* 10: 251–252.

Lachmann, L.M. (1943b) 'The Role of Expectations in Economics as a Social Science', *Economica* 10: 12–23.

Lachmann, L.M. (1950) 'Joseph A. Schumpeter, 1883–1950', *South African Journal of Economics* 18: 215–218.

Lachmann, L.M. (1951) 'The Science of Human Action', *Economica* 18: 412–427.

Lachmann, L.M. (1970) *The Legacy of Max Weber*, London: Heinemann.

Lachmann, L.M. (1973) 'Sir John Hicks as a Neo-Austrian', *South African Journal of Economics* 41: 195–207.

Lachmann, L.M (1976) 'From Mises to Shackle: An Essay on Austrian Economics and the Kaleidic Society', *Journal of Economic Literature* 14: 54–62.

Lachmann, L.M. (1977) *Capital, Expectations and the Market Process: Essays on the Theory of the Market Economy*, Kansas City: Sheed Andrews & McMeel.

Lachmann, L.M. (1982) 'Ludwig von Mises and the extension of subjectivism', in I.M. Kirzner (ed), *Method, Process, and Austrian Economics: Essays in Honor of Ludwig von Mises*, Lanham: Lexington Books, pp 31–40.

Lachmann, L.M. (1986) *The Market as an Economic Process*, Oxford: Blackwell Publishing.

Leff, G. (1975) *William of Ockham: The Metamorphosis of Scholastic Discourse*, Manchester: Manchester University Press.

Locke, J. (1975 [1690]) *An Essay Concerning Human Understanding*, Book I, edited by P.H. Nidditch, Oxford: Clarendon.

Mach, E. (1914) *The Analysis of Sensations*, Chicago: Court (translated from the 5th German edition, first German edition 1886).

Mach, E. (1907 [1883]) *The Science of Mechanics*, Chicago: Court (translated from 'Die Mechanik in ihrer Entwickelung', first published 1883).

Marshall, A. (1890) *Principles of Economics*, London: Macmillan.

McCloskey, D.N. (1983) 'The Rhetoric of Economics', *Journal of Economic Literature* 21(2): 481–517.

McCloskey, D.N. (1985) *The Rhetoric of Economics*, Madison: University of Wisconsin Press.

Menger, C. (1871) 'Grundsätze der Volkswirthschaftslehre', in *The Collected Works of Carl Menger*, Vol 1, LSE Reprints 1934.

Menger, C. (1883) 'Untersuchungen über die Methode der Socialwissenschaften und der Politischen Oekonomie insbesondere', in *The Collected Works of Carl Menger*, Vol 2, LSE Reprints 1933.

Menger, C. (1883/1960) *Problems of Economics and Sociology*, translation of *Menger 1883* by F.J. Nock, University of Illinois Press.

Menger, C. (1884) 'Die lrrthümer des Historismus in der deutschen Nationalökonomie', in *The Collected Works of Carl Menger*, Vol 3, LSE Reprints 1935, pp 1–97.

Menger, C. (1887) 'Zur Kritik der Politischen Oekonomie', in *The Collected Works of Carl Menger*, Vol 3, LSE Reprints 1935, pp 99–131.

Menger, C. (1889) 'Grundzüge einer Klassifikation der Wirtschaftswissenschaften', *The Collected Works of Carl Menger*, Vol 3, LSE Reprints 1935, pp 185–218.

Menger, C. (1933–6) *The Collected Works of Carl Menger*, London: London School of Economics (LSE) Reprint.

Menger, K. (1973) 'Austrian Marginalism and Mathematical Economics', in J.R. Hicks and W. Weber (eds), *Carl Menger and the Austrian School of Economics*, Oxford: Clarendon, pp 38–60.

Mill, John Stuart (1843) *A System of Logic, Ratiocinative and Inductive: Volume 1: Being a Connected View of the Principles of Evidence, and the Methods of Scientific Investigation*, London: John W. Parker.

Mill, J.S. (1863) *Utilitarianism*, London: Parker, Son & Bourn.

Mill, J.S. (1872 [1843]) *A System of Logic*, 9th edn, Vol II, London: Longmans.

Mises, L. (1931) 'Vom Weg der Subjektivistischen Wertlehre', in L. Mises and A. Spiethoff (eds), *Probleme der Wertlehre*, Munich and Leipzig: Duncker & Humblot, pp 73–94.

Mises, L. (1933) *Epistemological Problems of Economics*, Princeton: Van Nostrand.

Mises, L. (1949) *Human Action*, London: Hodge.

Mises, L. (1951) *Positivism*, Cambridge, MA: Harvard University Press.

Mises, L. (1958) *Theory and History*, London: Jonathan Cape.

Mises, L. (1962) *The Ultimate Foundations of Economic Science*, Princeton: van Nostrand.

Mises, L. (1978) *The Ultimate Foundation of Economic Science*, 2nd edn, Kansas City: Sheed Andrews & McMeel.

Mises, L. and Spiethoff, A. (eds) (1931) *Probleme der Wertlehre*, Munich and Leipzig: Duncker & Humblot.

Mittermaier, K. (2023) *A Realist Philosophy of Economics*, Bristol: Bristol University Press.

Mohanty, J.N. (1964) *Edmund Husserl's Theory of Meaning*, The Hague: Martinus Nijhoff.

Nagel, E.S.H. (1979) *The Structure of Science*, London: Routledge.

Nagel, E.S.H. and Samuelson, P.A. (1963) *American Economic Review*. Vol LIII, Papers and Proceedings, May 1963 pp 211–219, 229– 231 and 231.

Pareto, V. (1953 [1900]) 'On the Economic Phenomenon', *International Economic Papers* 3: 180–207.

Pareto, V. (1971 [1906]) *Manual of Political Economy*, translated from the French edition of 1927 by A. Schwier, London: Macmillan.

Phillips, D.L. (1977) *Wittgenstein and Scientific Knowledge: A Sociological Perspective*, London: Macmillan.

Poincare, H. (1914 [1908]) *Science and Method*, London: Nelson.

Polanyi, M. (1958) *Personal Knowledge: Towards a Post-Critical Philosophy*, London: Routledge.

Popper, K. (1959) *The Logic of Scientific Discovery*, London: Hutchinson (original German edition 1935).

Popper, K. (1945) *The Open Society and its Enemies*, London: Routledge.

Popper, K. (1972) *Objective Knowledge*, Oxford: Clarendon.

Quine, W.V. (1953) 'Two Dogmas of Empiricism', in *From a Logical Point of View*, Cambridge, MA: Harvard University Press.

Quine, W.V. (1960) *Word and Object*, Cambridge, MA: MIT Press.

Quine, W.V. (1969) *Ontological Relativity and Other Essays*, New York: Columbia University Press.

Ricardo, D. (1817/1965) *On the Principles of Political Economy and Taxation*, London: Dent.

Robbins, L. (1935 [1932]) *An Essay on the Nature and Significance of Economic Science*, 2nd edn, London: Macmillan.

Ross, J.J. (1970) *The Appeal to the Given: A Study in Epistemology*, London: Allen & Unwin.

Rothbard, M.N. (1976) 'Praxeology: The Methodology of Austrian Economics', in E.G. Dolan (ed), *The Foundations of Modern Austrian Economics*, Kansas City: Sheed & Ward, pp 19–39.

Ryle, G. (1949) *The Concept of Mind*, London: Hutchinson.

Schlick, M. (1974) *General Theory of Knowledge*, Vienna: Springer.

Schumpeter, J.A. (1952) *Ten Great Economists: From Marx to Keynes*, London: Allen & Unwin.

Schumpeter, J.A. (1954) *History of Economic Analysis*, New York: Oxford University Press.

Shackle, G.L.S. (1966) *The Nature of Economic Thought*, Cambridge: Cambridge University Press.

Shackle, G.L.S. (1969) *Decision, Order and Time in Human Affairs*, 2nd edn, Cambridge: Cambridge University Press.

Shackle, G.L.S. (1972) 'Marginalism: The Harvest', *History of Political Economy* 4: 587–602.

Shackle, G.L.S. (1972) *Epistemics and Economics*, Cambridge: Cambridge University Press.

Shackle, G.L.S. (1983) 'An Interview with G.L.S. Shackle', *Austrian Economic Newsletter* 4(1).

Smith, A. (1776) *An Inquiry into the Nature and Causes of the Wealth of Nations*, London: W. Strahan and T. Cadell.

Smith, A. (1892 [1759]) *Theory of Moral Sentiments*, London: Bell.

Smith, A. (1937 [1776]) *An Inquiry into the Nature and Causes of the Wealth of Nations*, Cannan edition, New York: Random House.

Sokolowski, R. (1964) *The Formation of Husserl's Concept of Constitution*, M. Nijhoff.

Spadaro, L.M. (ed) (1978) *New Directions in Austrian Economics*, Kansas City: Sheed Andrews & McMeel.

Streissler, E. (1973) 'To What Extent Was the Austrian School Marginalist?', in R. Black et al (eds), *The Marginal Revolution in Economics*, Durham: Duke University Press.

Tarascio, V.J. (1973) 'Vilfredo Pareto and Marginalism', in R.C.D. Black et al (eds), *The Marginal Revolution in Economics*, Durham: Duke University Press, pp 140–159.

Walras, L. (1873) 'Principe d'une théorie mathématique de l'échange', *Journal des Économistes* 34: 5–22.

Walras, L. (1874) *Elements of Pure Economics*, translated by W. Jaffé, London: Allen & Unwin.

Walras, L. (1954) *Elements of Pure Economics*, translated from the edition of 1926 by W. Jaffé, London: Allen & Unwin.

Weinberg, J.R. (1936) *An Examination of Logical Positivism*, London: Kegan Paul.

Whitehead, A.N. (1925) *Science and the Modern World*, New York: Macmillan Company.

Wieser, F. (1929) *Gesammelte Abhandlungen*, edited by F.A. Hayek, Tübingen: Mohr.

Winch, D. (1972) 'Marginalism and the Boundaries of Economic Science', *History of Political Economy* 4: 325–343.

Wittgenstein, L. (1922) *Tractatus Logico-Philosophicus*, New York: Harcourt, Brace and Company Inc.

Wittgenstein, L. (1953) *Philosophical Investigations*, translated by G. Anscombe, New York: Macmillan.

Wright, G.H. von (1975 [1957]) *The Logical Problem of Induction*, 2nd edn, Oxford: Blackwell.

Index

K

L

M